FM 21-150

US ARMY BASIC FIELD MANUAL

UNARMED DEFENSE FOR THE AMERICAN SOLDIER

1942

WORLD WAR II

CIVILIAN REFERENCE EDITION

UNABRIDGED MILITARY HANDBOOK ON PRINCIPLES OF HAND-TO-HAND COMBAT AND SELF-DEFENSE

U.S. WAR DEPARTMENT

Doublebit Press

www.doublebitpress.com | Cherry, IL, USA

Originally published in 1942 by the U.S. War Department.

A Part of the Military Manual Series

Doublebit Press Civilian Reference Edition ISBNs
Hardcover: 978-1643892184
Paperback: 978-1643892191

FM 21–150

BASIC FIELD MANUAL

UNARMED DEFENSE FOR THE AMERICAN SOLDIER

UNITED STATES
GOVERNMENT PRINTING OFFICE
WASHINGTON : 1942

WAR DEPARTMENT,
WASHINGTON, June 30, 1942.

FM 21–150, Unarmed Defense for the American Soldier, is published for the information and guidance of all concerned.

[A. G. 062.11 (5–3–42).]

BY ORDER OF THE SECRETARY OF WAR:

G. C. MARSHALL,
Chief of Staff.

OFFICIAL:

J. A. ULIO,
Major General,
The Adjutant General.

DISTRIBUTION:

D (5); R 1–10, 17 (5); Bn 7 (1), 11, 18 (5), 19 (10); C 7 (5), 19 (20).

(For explanation of symbols see FM 21–6.)

TABLE OF CONTENTS

BASIC FIELD MANUAL

UNARMED DEFENSE FOR THE AMERICAN SOLDIER

Section I

GENERAL

■ 1. Scope.—This manual describes a method of self-protection available to the American soldier, if through any circumstance he is unarmed or unable to use his weapons.

■ 2. Purpose of Training.—The object of this training is to develop the soldier in the art of unarmed self-defense, and to improve his skill in the use of his basic weapons, through speeded reflexes. Confidence in his own ability unarmed, like confidence in his weapons, makes a man a better soldier.

■ 3. Necessity for Training in Unarmed Defense.—The average soldier, if trained only in the use of his weapons, loses his effectiveness if these weapons are lost or fail to function. However, particularly in hand-to-hand fighting, if a soldier should be deprived of his weapon or have it destroyed, he is at the mercy of the enemy. This appears to apply mainly to the Infantry, and probably the greatest value of American unarmed defense will be to that arm. Nevertheless, in these days of fluid warfare, troops in rear echelons, artillery, and antitank units might find themselves in hand-to-hand combat with no defensive weapons except sidearms and bare hands.

■ 4. Training Program.—The training of the soldier in unarmed defense requires no special equipment or uniform. Clothing will depend upon the season of the year and the state of the weather. Work outdoors is preferable since a greater number of men can be trained simultaneously. Thirty minutes' instruction or practice each day will make a man adept in a very short period of time. If no additional time is available, this part of the training can be integrated into the physical training program. It is desirable, in order to obtain the maximum results, that the instruction follow closely the steps outlined in this manual. However, it is realized that all units will not have the time to go through the entire book. For units with a limited time allotment for this subject, it is recommended that the following be taught:

a. Section II.—Principles of unarmed defense.

b. Section III.

(1) One escape from underarm front body hold.
(2) One escape from front overarm body hold.
(3) One escape from rear underarm body hold.

c. Section V.

(1) One escape from two-handed front choke.
(2) One escape from two-handed rear choke.
(3) One escape from one-arm rear strangle.

d. Section VIII.

(1) One defense for downward stroke of knife.
(2) One defense for upward stroke of knife.
(3) One defense for downward sword cut.
(4) One defense for sword lunge.

e. Section IX.

(1) One defense for downward blow of club.
(2) One defense for side stroke of club.
(3) One defense for reverse stroke of club.

f. Section X.

(1) One defense for pistol in front, right or left hand.
(2) One defense for pistol in back, right or left hand.

g. Section XI.—Complete section.

For military police units with limited time, it is recommended that in addition to the above, sections VI, VII, IX, X, XII, and XIV be practiced in their entirety.

■ 5. BACKGROUND OF UNARMED DEFENSE.—The original name of the method described in this manual has been lost in antiquity, but the art was developed by Chinese monks approximately in the twelfth century. The monastic rules forbade the monks to use weapons, but as they were constantly being attacked by nomads and robber bands, they had to devise a weaponless defense, utilizing only the skill of their bodies and the quickness of their brains. Through long experiment, trial and error, and loss of life they developed a means of defense that has remained basically unchanged through centuries. Late in the twelfth century, the Japanese became aware of this art and, characteristically, they copied it and claimed it as their own. They named this art "Jiu Jitsu," and established a genealogy for it which they claimed extended back to their mythological age. The Jiu means "gentle" and Jitsu means "art" or "practice." Therefore, Jiu Jitsu is "the gentle art." The systems taught were multitudinous and varied until the year 1882 when Professor Jigoro Kano, a man who had studied all the better systems, established the Kodokan, "a school for studying the way" and called his system "Judo." This name means "the way, or principle." This school, with its roots in Tokyo, sent out branches throughout the civilized world. One branch, founded in 1921, had its headquarters in New York. It was called "The New York Dojo," and while catering mainly to Japanese, admitted Occidentals who were interested. However, progress of the Occidentals was slow, due to the fact that their instruction was mainly in competitive work. The holds were ineffective because the correct principles were not taught. Very little of the defensive or protective tactics was taught. Since this was the type of Judo in which the average American was interested, he soon dropped out of the school. A group of young Americans, disgusted with this procedure, set out to develop a system of self-defense suited to the American temperament and needs. They called their organization "The American Judo Club" and dedicated themselves to removing Oriental terminology from the new system. They produced as good a system as the Japanese and far outstripped it in the effectiveness of method. With a knowledge of American unarmed defense the American soldier will be equipped to meet the Judo men in the game which they have chosen to claim as their own.

■ 6. **Methods of Training.**—*a*. Regulation physical training formations may be used for practice (see FM 21–20). From the extended formation of four columns have the first and second columns face each other and the third and fourth columns face each other. Each man will then have a partner with whom to practice. Special note should be taken that the even-numbered men *do not uncover.* The above formation applies to a unit the size of a platoon or larger. Any unit smaller than a platoon should be formed in a column of twos and then have the columns face each other. It is recommended that when working throwing tricks, twice the normal distance be taken.

b. The instructor will explain the attack and demonstrate the proper defense on a competent assistant, executing the movement rapidly to show its effectiveness. The defense is then executed again, as near slow motion as possible with an accompanying explanation. The attacking squads and the opposing defending squads are then designated, possibly using the letter "a" for attackers and letter "b" for defenders. At a given signal the attackers move to the attack and the defenders attempt to work the proper defense while the assistant instructors make corrections. Emphasis should be placed on precision first. Speed can be developed later. Most of the defenses are equally effective on either side. When two defending squads have mastered the defense, the situation is reversed and the defending squads become attackers. Progress to a new trick is made only when the students have demonstrated a working knowledge of the previous one. No more than three tricks should be taught in any 30-minute period, as confusion would result. Encourage the men to practice in their spare time, emphasizing that proficiency in unarmed defense is predicated on repetition until a movement becomes almost instinctive. It is not difficult to arouse the interest of the men in this subject, since the desire to excel physically is a characteristic of the average American. Since even the smallest can be shown that his lack of size is no handicap, there will be no difficulty arising from indifference. The main problem will be to keep enthusiasts from trying more tricks than they can possibly assimilate. Another point that should be emphasized is the desirability of eliminating the stigma of the so-called "foul tactic" which is usually ascribed to unarmed defense. It might be well to point out that an individual who attacks with a club, knife, gun, or any other weapon is not subscribing to any recognized rules of combat. In hand-to-hand combat, there are no referees, no judges, and no timekeeper. You are on your own. No measure of defense is too extreme when your life is in danger. The defenses in this manual might be the means of saving your life or the life of a comrade.

SECTION II

BASIC PRINCIPLES

■ 7. **PRINCIPLES OF UNARMED DEFENSE.**—It is of the utmost importance that in order to learn unarmed defense in the manner necessary to use it effectively, the principles be first mastered.

a. The first principle is "balance." The accompanying illustrations will be used to point out the principle of balance. Figure 1 ① illustrates a man in balance with his legs apart. It is shown that the individual in this position is on balance from right to left and from left to right. However, figure 1 ② shows that he is definitely off balance forward, and figure 1 ③ shows that he is also off balance backward. It is usually assumed that when an individual assumes the position of the charge he is on balance all the way around. Figure 2 ① shows that in a position of charge with the right foot forward, the individual is on balance from right front to left rear and from left rear to right front. However, figure 2 ② shows that in this position he is off balance to the right rear, and figure 2 ③ shows that he is off balance to the left front. Notice the small amount of effort required to take an individual off balance when you know in what directions he is strong and in what directions he is weak. No matter what position an individual assumes, he is off balance in some direction.

b. The second principle is "use of the internal oblique muscles." These muscles have been named by the Japanese Shita-hara, pronounced "stahara." The internal oblique muscles are located in the center of the body in the lower abdomen between the hip bones. The power for every defense must come or be centered in these muscles.

c. Principle three is the ability to utilize an opponent's momentum or an opponent's strength to bring about his downfall. You always assume that your opponent is stronger than you are and never attempt to oppose him directly, but rather utilize his impetus or momentum to carry him on his

FIGURE 1 ①.

way, the difference being that you direct the movement.

d. Principle four is to attack your opponent on the spot

where he is weakest with the greatest amount of power that you can concentrate on that one point. The axiom of this principle is "My maximum strength against your minimum."

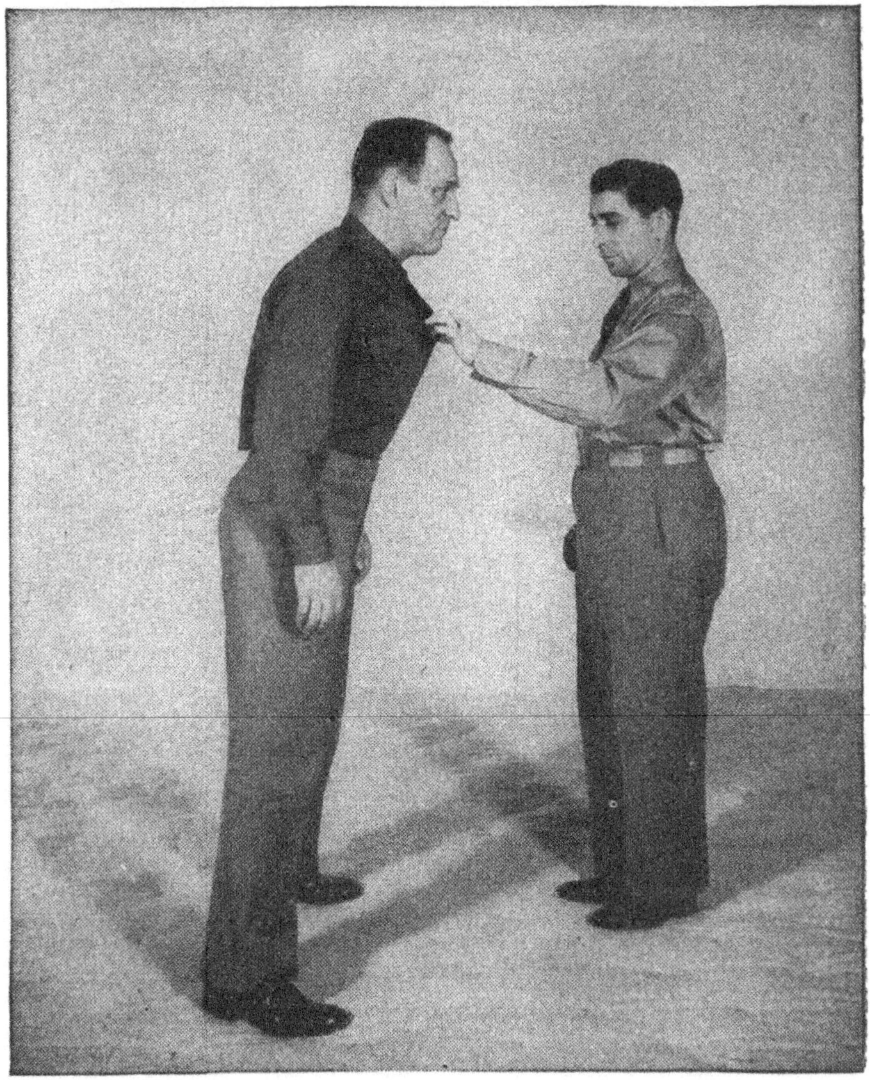

FIGURE 1 ②.

To illustrate this point, if an opponent were to grasp your wrist, instead of trying to tear your wrist from between his fingers by main strength of your arms, you would attempt to concentrate the power of your legs, body, and arms against his fingers.

FIGURE 1 ③.

e. The fifth and last principle is a knowledge of "the major and minor operations." This knowledge is essential to prevent injury while practicing. "The major operation" means either getting out of danger or getting the essential part of a hold. "The minor operation" is the application of the pressure. If both of these were run together, the results

FIGURE 2 ①

might be a broken bone or other serious injury to a partner. Therefore, in practice, be judicious and apply the "major and minor operations" separately.

FIGURE 2 ②

FIGURE 2 ③

Section III

WRIST ESCAPES

■ 8. **Defense Against Two-Hand Overhand Grip on Both**

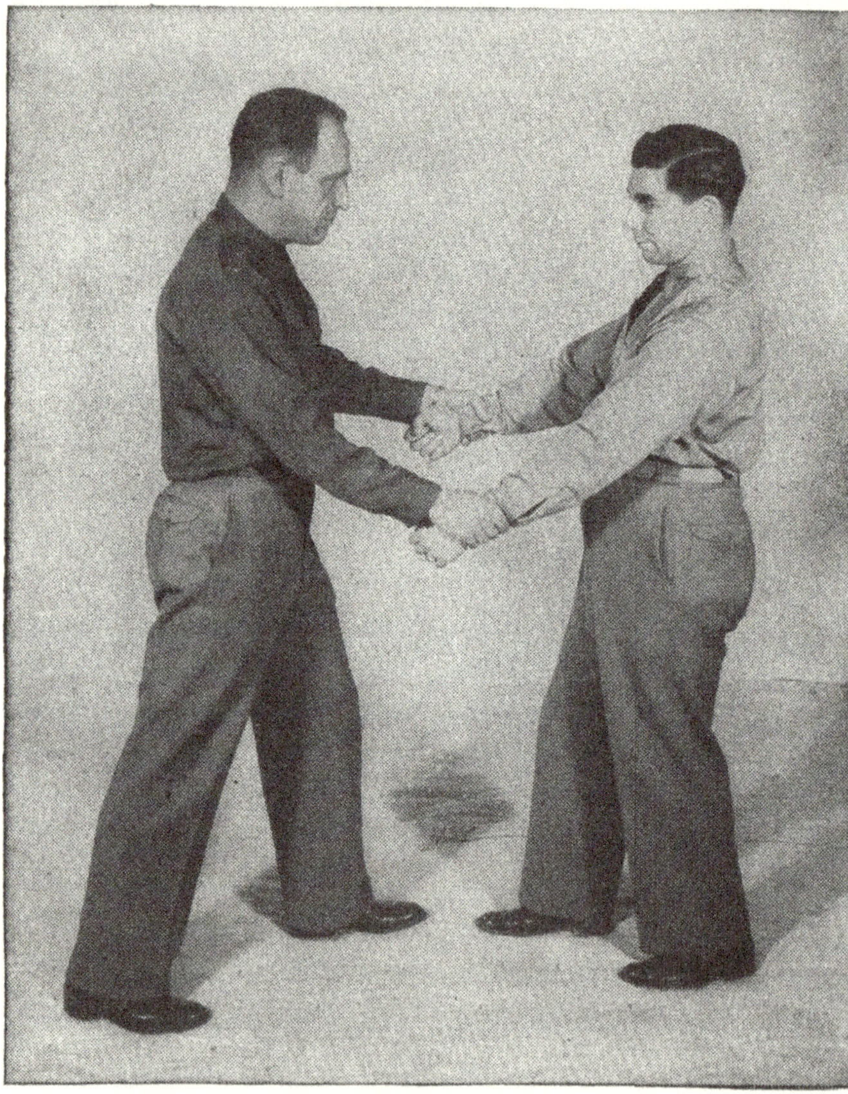

Figure 3 ①.

Wrists (Simple).—You are grasped by the wrists in the manner illustrated in figure 3 ①. You immediately step forward with either foot, in this case (fig. 3 ②) the right one, at the same time bending the arms so that the elbows are close to the lower abdomen. The escape is accomplished by

FIGURE 3 ②.

straightening the legs, pulling back with the body, and pushing the arms upward in one motion (fig. 3 ③). The faster this defense is worked, the more effective the escape.

■ 9. **Defense Against Two-Hand Overhand Grip on Both Wrists (Advanced).**—Figure 3 ① is the original hold. Figure 4 ① illustrates the first movement, in which your right hand

Figure 3 ③.

reaches across and grasps your opponent's right wrist. Following this movement, the left wrist is pulled out of the grasp of your opponent's right hand by pushing with your right hand and pulling with your left arm. Figure 4 ② illustrates that you have reached across with your freed left hand and grasped your opponent's left wrist with your thumb up. By

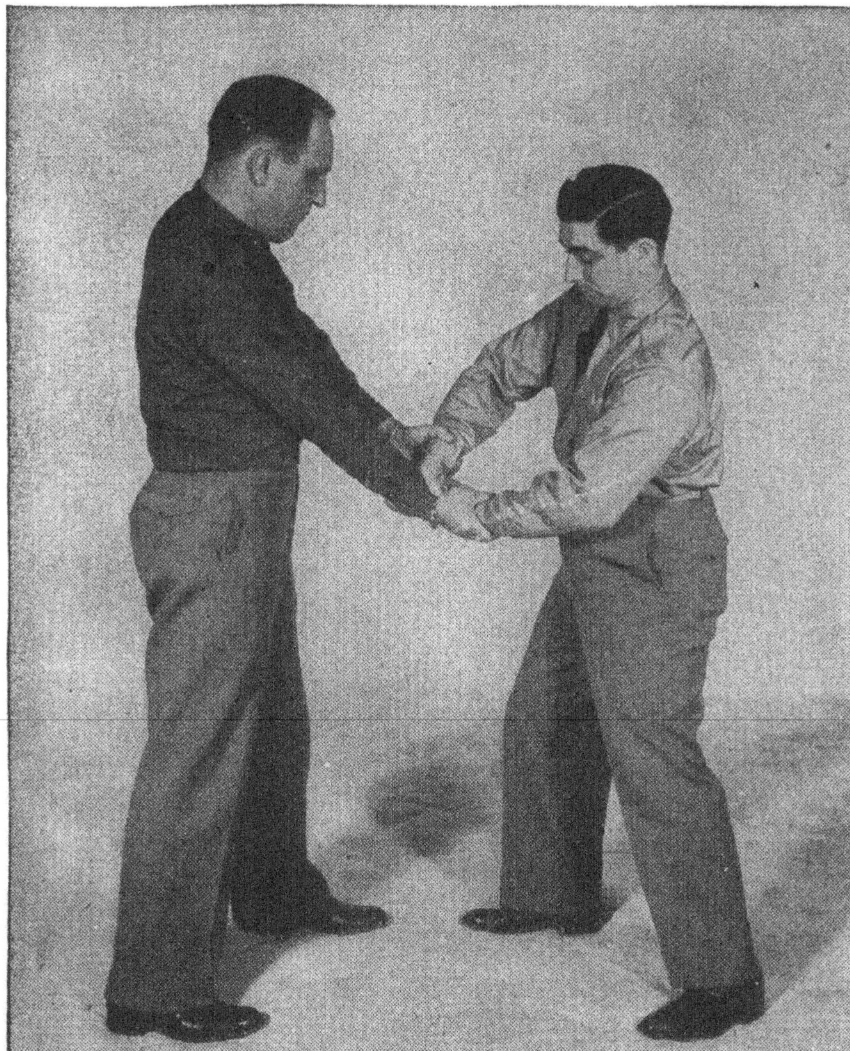

FIGURE 4 ①.

pushing on his left wrist, you will find no difficulty in releasing his grip on your right wrist. Figure 4 ③ illustrates the completion of the defense, showing that you have lifted your opponent's left arm over his own right forearm and have him in a position where you will have no difficulty in snapping his left elbow.

FIGURE 4 ②.

■ 10. DEFENSE AGAINST TWO-HAND UNDERHAND GRIP ON BOTH WRISTS.—Figure 5 ① illustrates the original hold. This defense is just about the reverse of that for the two-hand overhand grip on both wrists. The elbows are again brought close to the lower abdomen and a step forward is again taken as illustrated in figure 5 ①. Then, turning the head to one

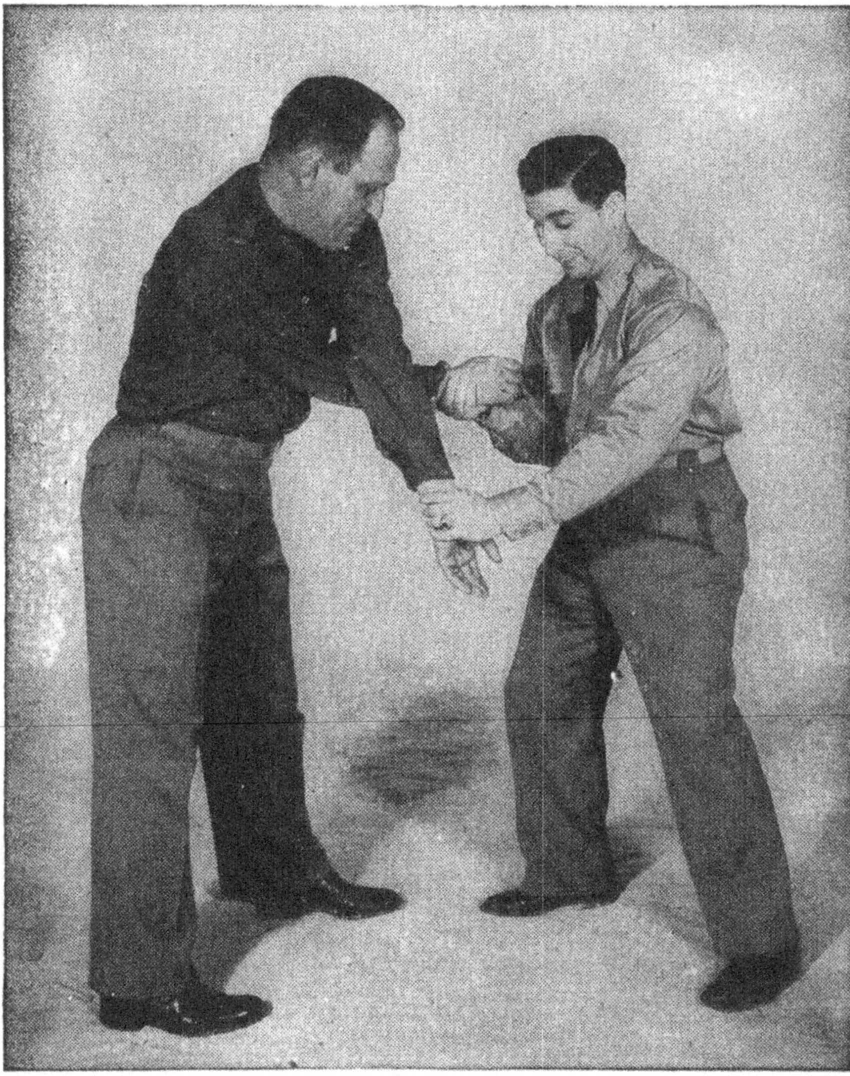

FIGURE 4 ③.

side to avoid bumping heads with a hard-headed opponent (fig. 5 ②), the body is bent swiftly from the waist. At the same time, pressure is brought to bear on your opponent's thumb, accomplishing the escape as in figure 5 ③.

■ 11. **Defense Against Left-Hand Grip on Right Wrist, or Right-Hand Grip on Left Wrist (Simple).**—Figure 6 ① illus-

Figure 5 ①.

trates the original hold. Your attacker has grasped your right wrist with his left hand. Your immediate action is to pull your left elbow close to your internal oblique muscles (see par. 7b), step forward with your right foot, and by *pushing* with your body, attempt to touch your right elbow against

FIGURE 5 ②.

your opponent's left elbow, at the same time turning the palm of your hand toward the floor as illustrated in figure 6 ②. Figure 6 ③ illustrates the completion of the escape, showing that your right hand is then in position to be brought smartly across your opponent's neck.

■ 12. DEFENSE AGAINST LEFT-HAND GRIP ON RIGHT WRIST, OR

FIGURE 5 ③.

RIGHT-HAND GRIP ON LEFT WRIST (ADVANCED).—The original hold is the same as in figure 6 ①. This time, however, your defense will be such that you can take your opponent prisoner. Reach across with your left hand and grasp your attacker's left wrist as illustrated in figure 7 ①. Then, by bringing

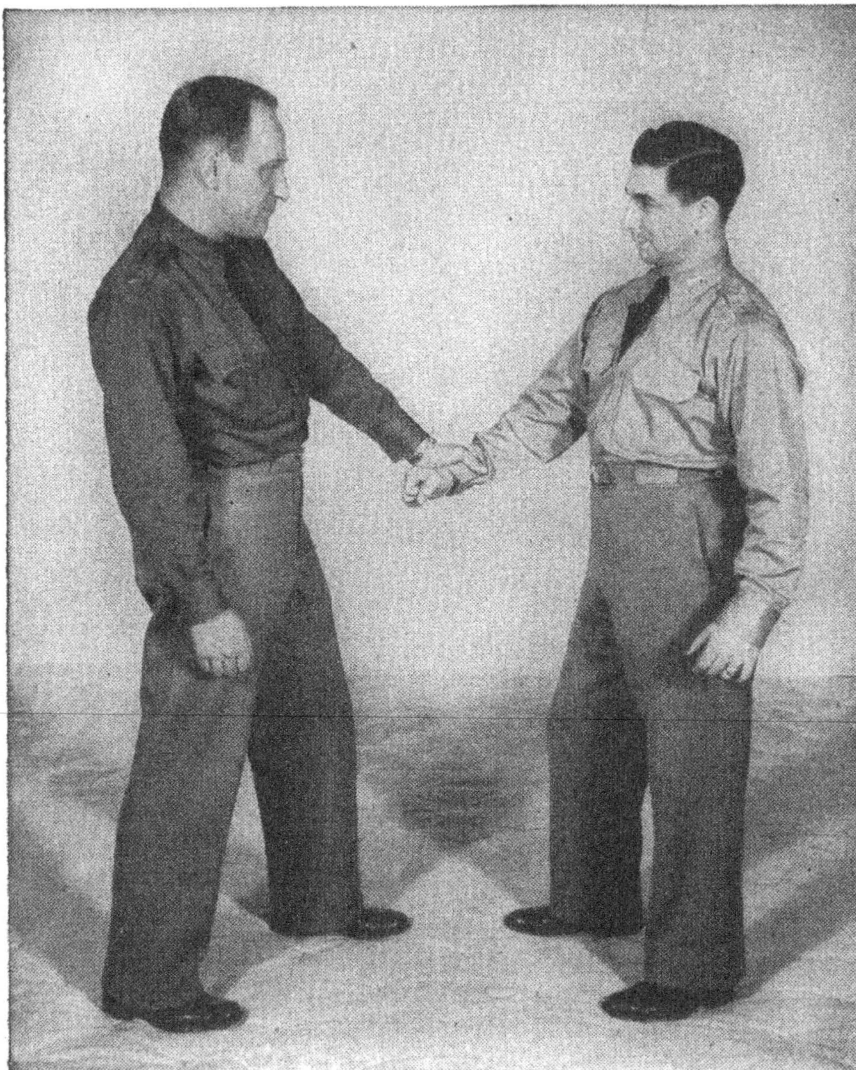

FIGURE 6 ①.

your elbows as close to your shita-hara as possible, you will turn your opponent's left hand to your left, applying pressure against his thumb as illustrated in figure 7 ②. As you turn the wrist, the thumb of your left hand is placed in the center of the knuckles on the back of your opponent's hand. As soon as your opponent has been forced to turn his back par-

FIGURE 6 ②.

tially to you, you will then reinforce the hold with your left hand with an identical one with your right hand. This will bring you to the position illustrated in figure 7 ③. The close-up in figure 7 ④ shows you the proper hold on your opponent's hand. The pressure, when it is necessary, will be applied toward your opponent's forearm rather than to either side.

FIGURE 6 ③.

Your elbows will be kept close to your side. All pressure will emanate from the shita-hara. You can now march your opponent anywhere you see fit. He is your prisoner.

■ 13. **Defense Against Two-Hand Grip on One Wrist.**—Figure 8 ① illustrates your opponent grasping your right wrist with both his hands. Now step forward with your

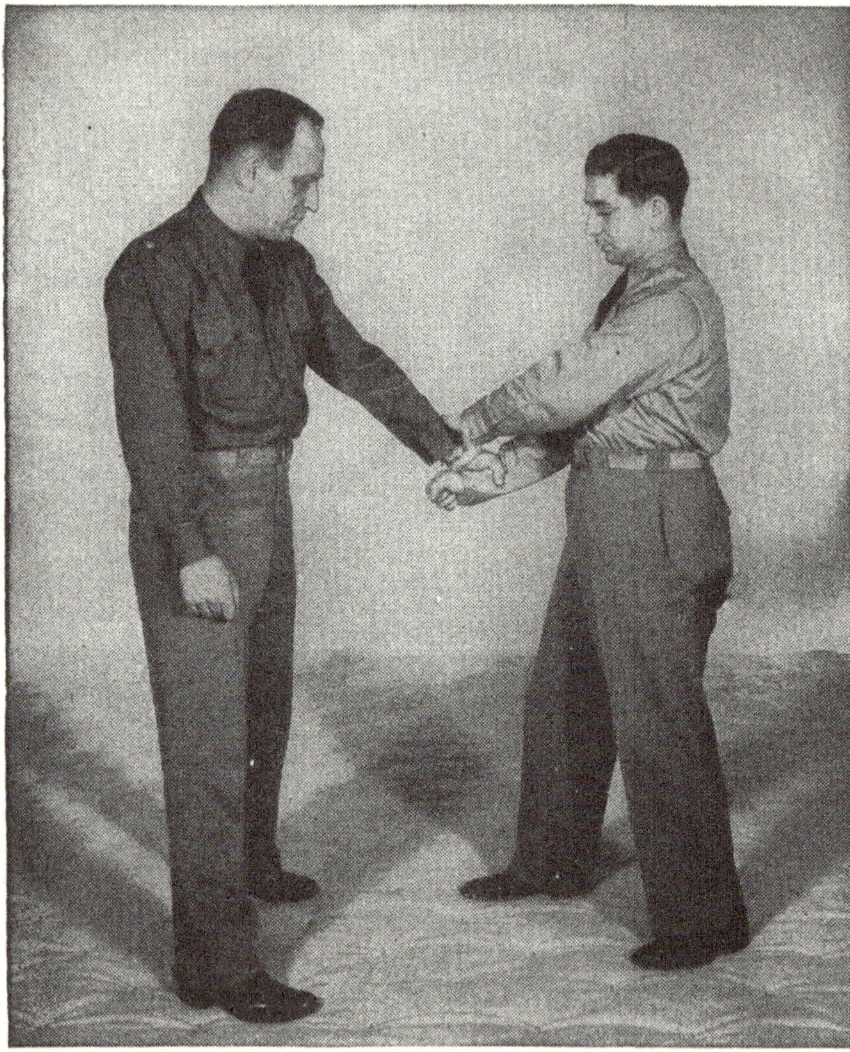

Figure 7 ①

right leg, bending both knees, body upright, and bring your right elbow close to your shita-hara. Your left hand will reach across and grasp your right fist. The position is illustrated in figure 8 ②. By straightening your legs, pulling back with the power of your body and of your arms, pressure is brought to bear on your opponent's thumbs, forcing him to release his hold. Figure 8 ③ illustrates the completion of the

FIGURE 7 ②.

escape, placing you in position to retaliate with a backhand with the edge of your right hand to your opponent's jaw or neck.

■ 14. Defense Against Right-Hand Grip on Right Wrist, or Left-Hand Grip on Left Wrist.—In the case illustrated by figure 9 ①, your opponent has grasped your right wrist with his right hand. Your immediate reaction is to grasp his right

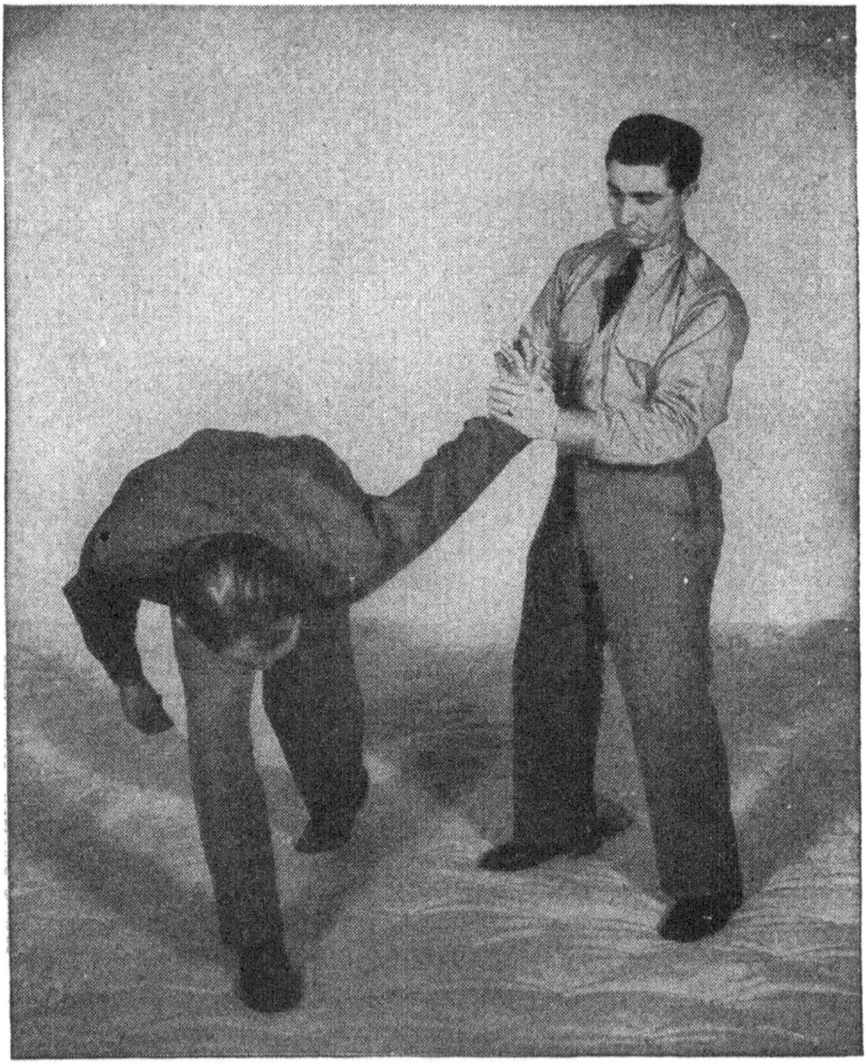

Figure 7 ⑧.

wrist with your right hand, taking a long step to his right rear with your left foot, pulling his right arm underneath your left arm as illustrated in figure 9 ②. Wrapping your upper arm over your opponent's upper arm and bringing your forearm or wrist underneath a spot about 1 inch above his elbow, you will then clamp your left hand on your chest as illustrated

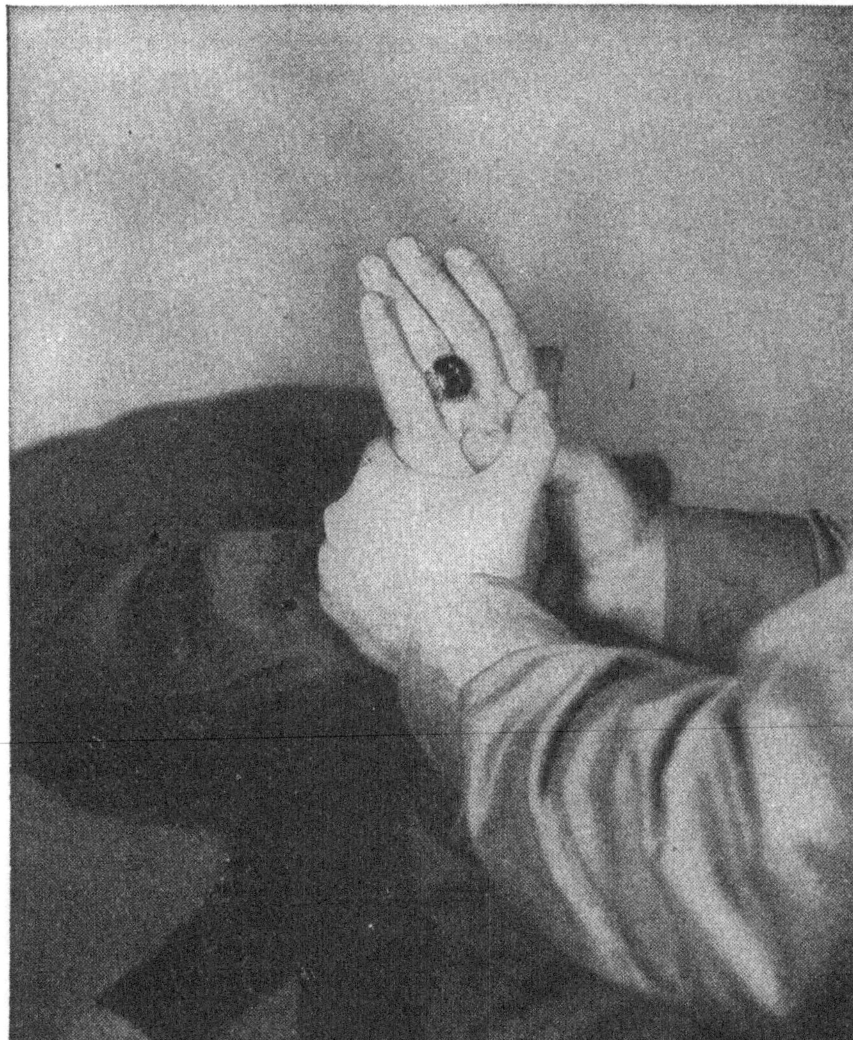

FIGURE 7 ④.

in figure 9 ③. You are now in position to bring pressure on either your opponent's elbow or his ulna nerve by pulling up with your left forearm and pushing down with your right arm. This will force your opponent to his toes and give you complete control of the situation. Care should be taken in working this defense, since it is very easy to break an arm utilizing this procedure.

FIGURE 8 ①.

■ 15. DEFENSE AGAINST LEFT-HAND GRIP ON RIGHT WRIST, OR RIGHT-HAND GRIP ON LEFT WRIST (FINGERS UPPERMOST).—Figure 10 ① illustrates the original hold, showing your opponent grasping your right wrist overhand with his left hand. Your immediate reaction is to reach across with your left hand and grasp his wrist with your thumb uppermost, at the

FIGURE 8 ②.

same time stepping forward with your *right* foot as illustrated in figure 10 ②. You then prevent him from releasing the grip on your wrist by pulling with your left hand, keeping it close to your right. You then endeavor to place your right elbow on your opponent's left elbow by pushing with the body and arm and continuing the pressure. This will bring him to the position illustrated in figure 10 ③ or completely to the

FIGURE 8 ③.

ground. If done rapidly, it can also result in a broken arm.

■ 16. DEFENSE AGAINST RIGHT-HAND GRIP ON RIGHT WRIST, OR LEFT-HAND GRIP ON LEFT WRIST (FINGERS UPPERMOST).—

This, on the surface, appears to be the same hold as in figure 10 ①, but it is not, since in one case the grip is with the hand on the same side, and in this case the grip is from

FIGURE 9 ①.

the hand on the opposite side as illustrated in figure 11 ①, where your opponent is grasping your right wrist with his right hand. Your first action is to place your right hand on your right biceps and your left forearm over the back of your opponent's hand, preventing him from releasing his grip, as illustrated in figure 11 ②. Continue the movement of your

FIGURE 9 ②.

left hand to the right, bringing it over the back of his right hand, under your right forearm, and locking it on your right biceps as illustrated in figure 11 ③. Holding him close to your chest, it is a simple matter to bend from the waist and thus force your opponent to his knees.

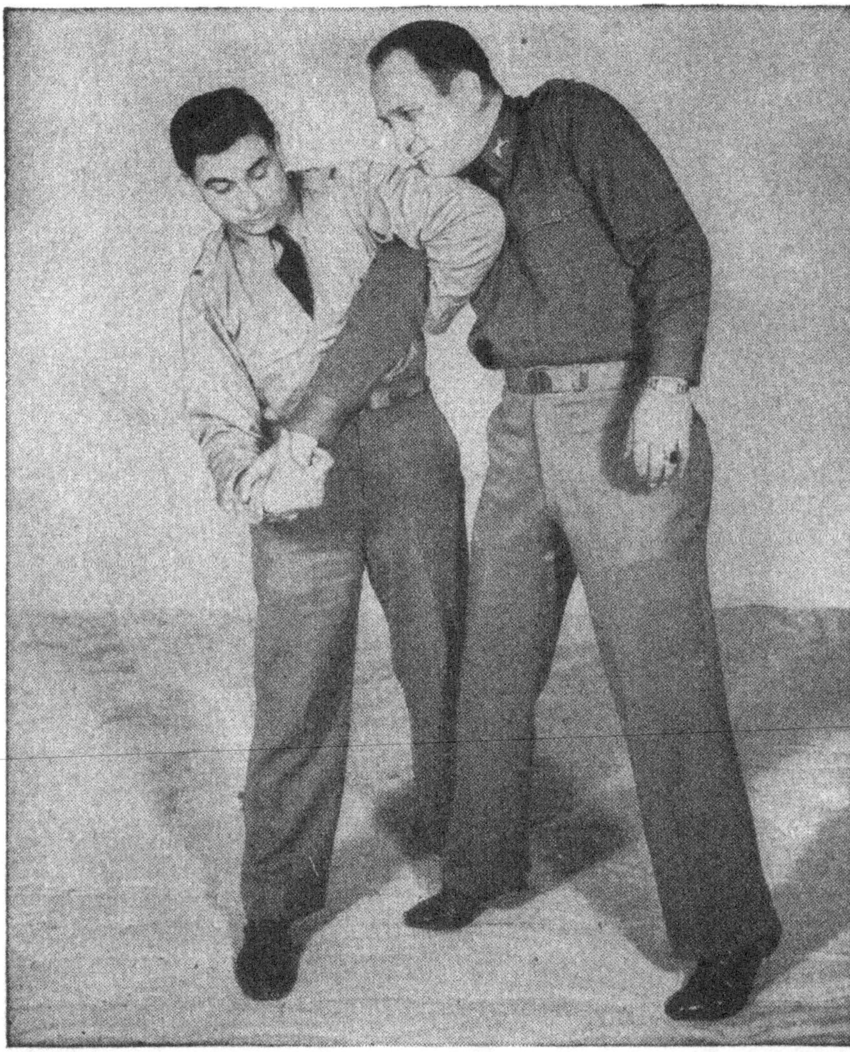

FIGURE 9 ③.

FIGURE 10 ①.

FIGURE 10 ②.

FIGURE 10 ③.

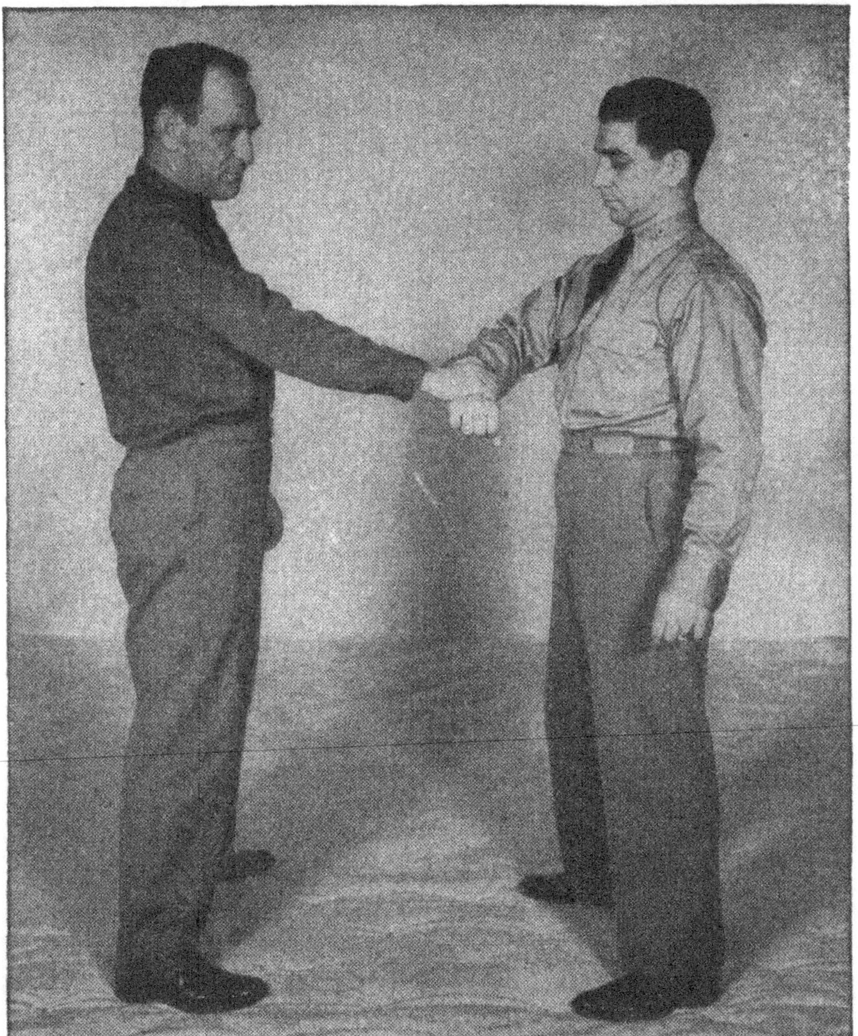

FIGURE 11 ①.

FIGURE 11 ②.

FIGURE 11 ③.

SECTION IV

ESCAPES FROM BODY HOLDS

■ 17. TWO ESCAPES FROM BEAR HUG.—Figure 12 ① illustrates

FIGURE 12 ①.

the bear hug. Your opponent has grasped you firmly around the waist from the front, and by powerful pressure of his arms is attempting to bend you over backward. A simple method of causing your opponent to release his hold is illustrated in figure 12 ②. Your fist is closed and your thumb is placed underneath the base of your opponent's nose. Pres-

FIGURE 12 ②.

sure on this very tender spot will cause your opponent to either pull his head backward or release his hold. Your right arm in this case is around your opponent's waist. By pushing with your thumb and pulling in on his waist, you can bear him over backward. Figure 12 ③ illustrates a more effective defense for the same hold. A thumb is brought to the jaw line on either side of your opponent's face and

FIGURE 12 ③.

running up the jaw line, brought to bear underneath the ear lobes. Pressure is upward and inward. This is an extremely tender spot and even the strongest of men will be quickly forced to release any hold.

■ 18. **Escape From Front Overarm Body Hold.**—Figure 13 ① illustrates that your oponent has grasped you around the body, including both your arms, so that you cannot use the

Figure 13 ①.

escapes explained by figure 12 ② and ③. Your first reaction is to bring your thumbs strongly into your opponent's groin, forcing his hips backward as illustrated in figure 13 ②. This leaves a space between your hips and his. Now pivot your hips on your left foot without moving it from the original spot, placing your right foot on the outside of your opponent's right foot, with the toe pointing in the same

FIGURE 13 ②.

direction as his. Your right arm slips up under his left armpit and grasps him anywhere in back. Your left hand grasps his upper arm, pulling strongly as illustrated in figure 13 ③. You now strike him strongly in the middle with your hips, at the same time twisting to the left, lifting with your right arm and pulling with your left hand. This will propel your opponent over your hips and to the ground with very little effort.

FIGURE 13 ③.

■ 19. ESCAPE FROM CHIN SHOVE.—A dangerous individual and possibly a powerful one is the one who attempts the attack illustrated in figure 14 ①. He grasps you firmly around the waist with his left hand and pushes strongly under your chin with his right hand, attempting to break your neck.

FIGURE 13 ④.

Your initial actions must be simultaneous. Push his right wrist upward with your left hand at the same time that you pull your head backward, pushing his hand and arm up in the manner illustrated in figure 14 ②. Now pivot on your left foot to the left, keeping your left foot in place, placing your right foot on the outside of your opponent's right foot

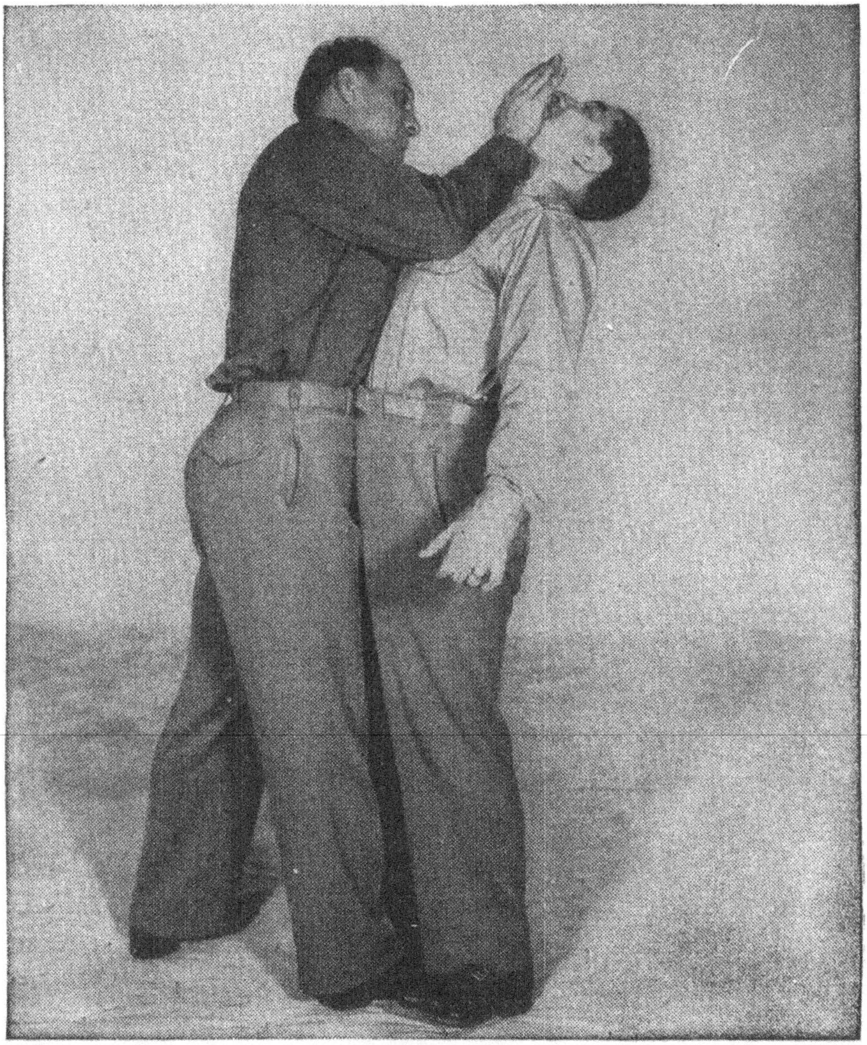

FIGURE 14 ①.

and pointing in the same direction. At the same time, you slip your right arm over his left shoulder and grasp his right shoulder blade as illustrated in figure 14 ③. You are now in position to throw him over your hips, as shown in figure 13 ④.

■ 20. **Escape From First Rear Underarm Body Hold.**—The attack illustrated in figure 15 ① shows your opponent behind you, grasping you around the waist and with his head hidden

Figure 14 ②.

directly behind yours. Your immediate reaction is to bring your head backward strongly, striking him in the face with the back of your head as illustrated in figure 15 ②. Your next movement is to follow up by bringing both elbows shoulder high and pivoting swiftly from left to right and from right to left as illustrated in figure 15 ③. It is impossible for you to miss striking him somewhere in the head

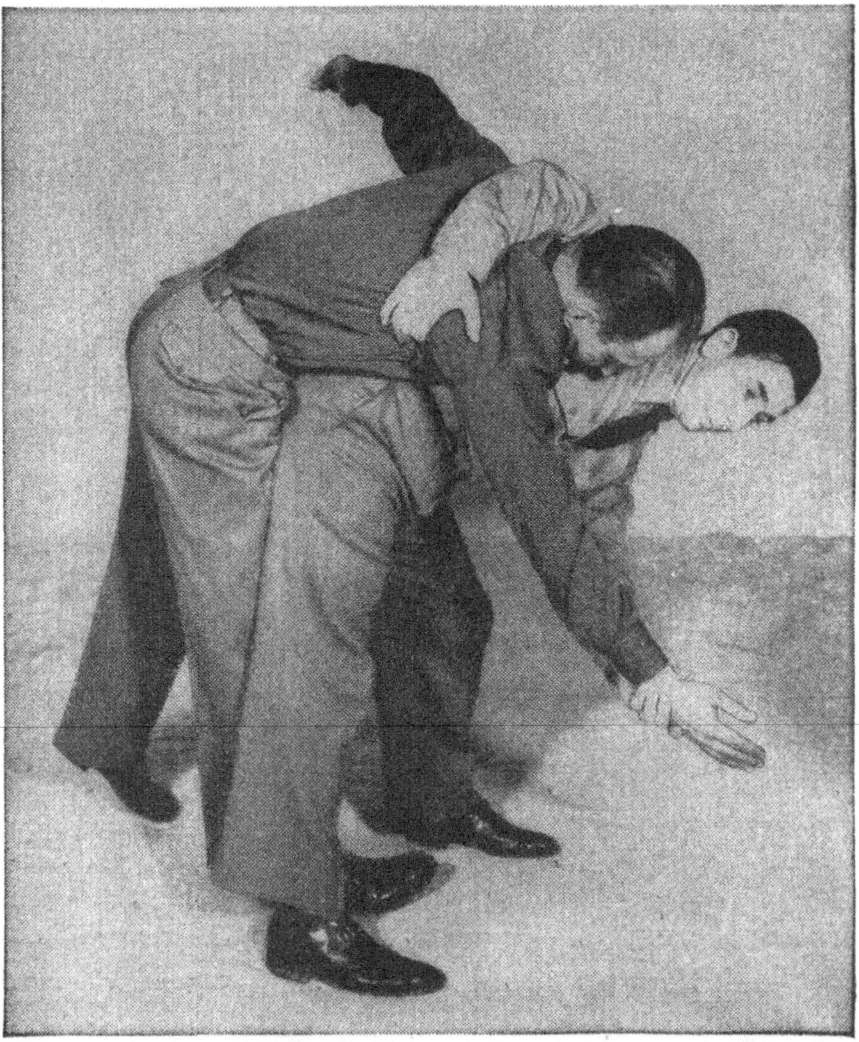

FIGURE 14 ③.

by means of this movement. On being struck in this manner, the results are usually devastating to your attacker.

■ 21. ESCAPE FROM SECOND REAR UNDERARM BODY HOLD.—Opponent grasps you as illustrated in figure 16 ①. Reach down with your left hand, placing it just above your opponent's left knee. Most of your weight will be borne on your left arm and his left leg as illustrated in figure 16 ②. By rest-

FIGURE 15 ①.

ing your weight in this manner, you then have a firm basis on which to pivot when you lift both feet from the ground and throw your left leg behind your opponent's right one.

As soon as your feet are firmly planted on the ground, you bring your left hand under your opponent's left knee joint and your right hand under your opponent's right knee joint,

FIGURE 15 ②.

figure 16 ③. By lifting and leaning backward, you can now easily overbalance your opponent over your left knee (fig. 16 ④).

■ 22. **Escape From Third Rear Underarm Body Hold.**—Your opponent uses the same grasp around the waist, under the arms from the rear, but this time braces himself by placing

Figure 15 ③.

one leg between yours and getting his head out of reach of your arms by placing it behind your shoulder blade (fig. 17 ①). Your immediate action is to bend swiftly from the waist, arms extended, and grasp your opponent's ankle (fig.

17 ②). Keeping your grasp on the ankle, you now straighten your body. This brings pressure to bear on your opponent's knee, causing him to release his hold and drop on his back

FIGURE 16 ①.

(fig. 17 ③). If your opponent should retain his hold, you merely fall backward on top of him, sitting on his midsection with great force.

■ 23. **Escape From Fourth Underarm Rear Body Hold.**—The hold around the waist from the rear is the same, but this time your opponent clasps his hands together. Your initial movement is to use the base of the thumb of either one of

Figure 16 ②.

your hands, pushing up on one of his index fingers (fig. 18 ①). This will cause him to release his hold. In this illustration, you use your left hand to lift the index finger of his right hand. Immediately the hold is loosened, grasp the back of your opponent's right hand with your right thumb, the fingers of your right hand clasping the palm of his hand around the little finger edge (fig. 18 ②). By bringing your

FIGURE 16 ③.

elbows close to your midsection and turning to the right, you will cause your opponent to turn his back to you due to the pressure brought on his wrist. During this turn, you will reinforce your original grasp with your right hand with an identical grasp with your left (fig. 7 ④). Terminating in the position illustrated in figure 18 ③, your elbows are close to your body and you maintain control of your opponent by

FIGURE 16 ④.

pressure of your thumbs against the back of his hand. In this position, you can march your opponent anywhere you please.

■ 24. ESCAPE FROM FIFTH UNDERARM REAR BODY HOLD.—When

your opponent grasps you around the waist this time, he grasps his own wrist. In this illustration, he has grasped his right wrist with his left hand. Your initial move is to

FIGURE 17 ①.

lock his right elbow with your left elbow joint and push on the back of his right hand with your left hand, attempting to force his fingers to touch his wrist (fig. 19 ①). By continuing pressure against the hand, you force your opponent to release his hold and by turning your body, you can take him into several of the come-alongs which will be explained later (fig. 19 ②). Extreme pressure can cause the dislocation of the wrist, or even a broken wrist.

FIGURE 17 ②.

■ 25. ESCAPE FROM FIRST OVERARM REAR BODY HOLD.—Your opponent grasps you from the rear and over the arms tightly (fig. 20 ①). Your initial movement is to cause him to loosen his hold even momentarily by either stepping on his instep or kicking him in the shin with your heel. The moment you feel the hold loosen, lower your body by bending your knees,

FIGURE 17 ③.

and, at the same time, raise your elbows to shoulder height (fig. 20 ②). From this position swing your elbows backward alternately; swing with the power of the shoulders and midsection, your elbows striking your opponent in either the short ribs or the solar plexus (fig. 20 ③). The first blow is usually a knock-out but very seldom can your opponent release his hold entirely before being struck two or three times.

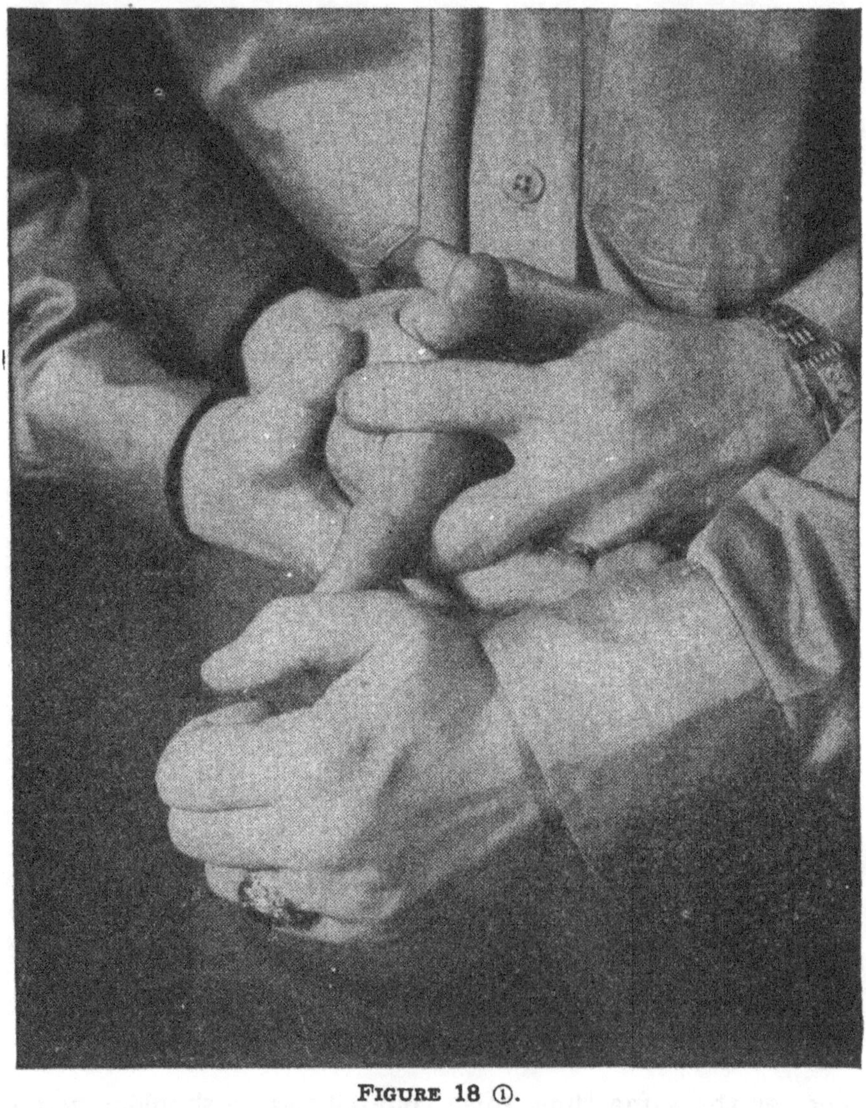

FIGURE 18 ①.

■ 26. ESCAPE FROM SECOND OVERARM REAR BODY HOLD.—The original attack by your opponent is again over the arms from the rear. Your initial movement is the same, loosening the grip by means of stepping on the instep or kicking the shins and raising the elbows shoulder high, lowering the body

FIGURE 18 ②.

simultaneously by bending the knees (fig. 20②). Then reach up with your right hand, grasping your opponent's right upper arm just above the elbow, your left hand grasping his right wrist at the same time that you move your right foot on a line with his right foot and on the outside of it (fig. 21 ①). From this position, strike backward with your hips against

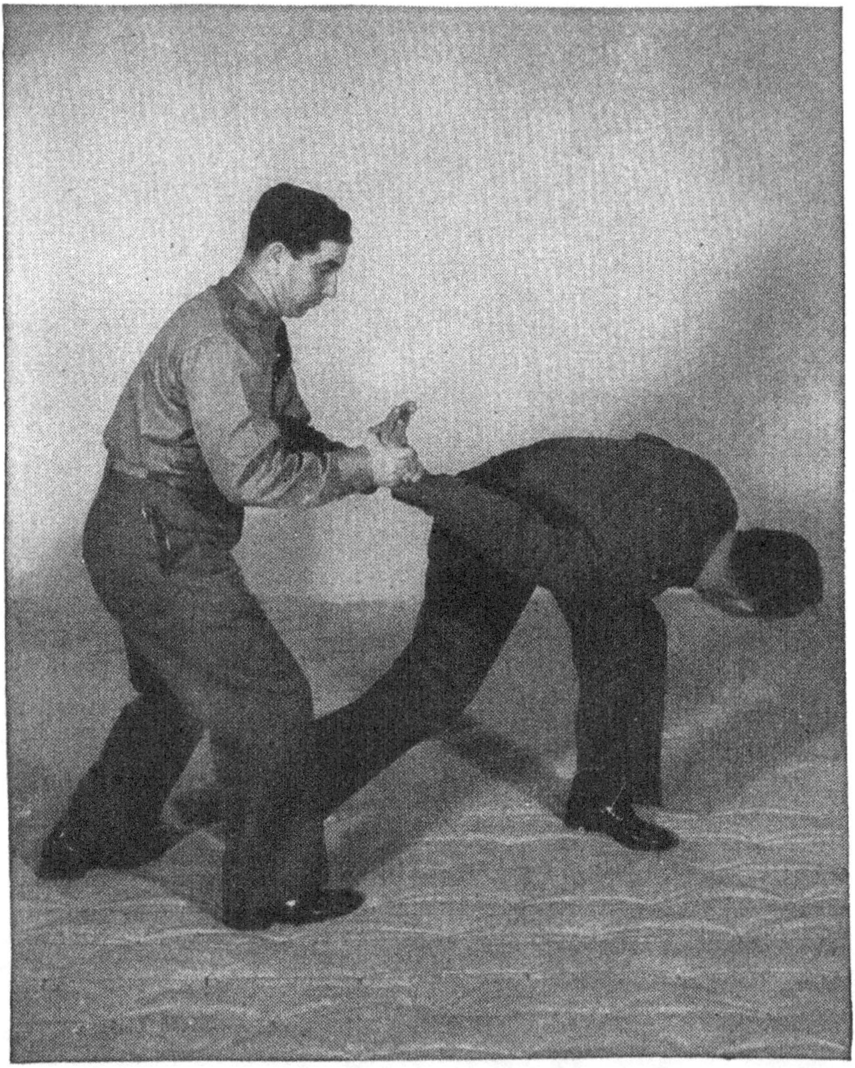

FIGURE 18 ③.

his midsection, at the same time bending swiftly from the waist, retaining your grip on his right arm (fig. 21 ②). Your opponent will fly over your head, striking the ground on his back.

FIGURE 19 ①.

FIGURE 19 ②.

FIGURE 20 ①.

FIGURE 20 ②.

FIGURE 20 ③.

FIGURE 21 ①.

FIGURE 21 ②.

SECTION V

DEFENSES AGAINST CHOKE HOLDS

■ 27. **FIRST DEFENSE AGAINST TWO-HAND FRONT CHOKE.**—Your

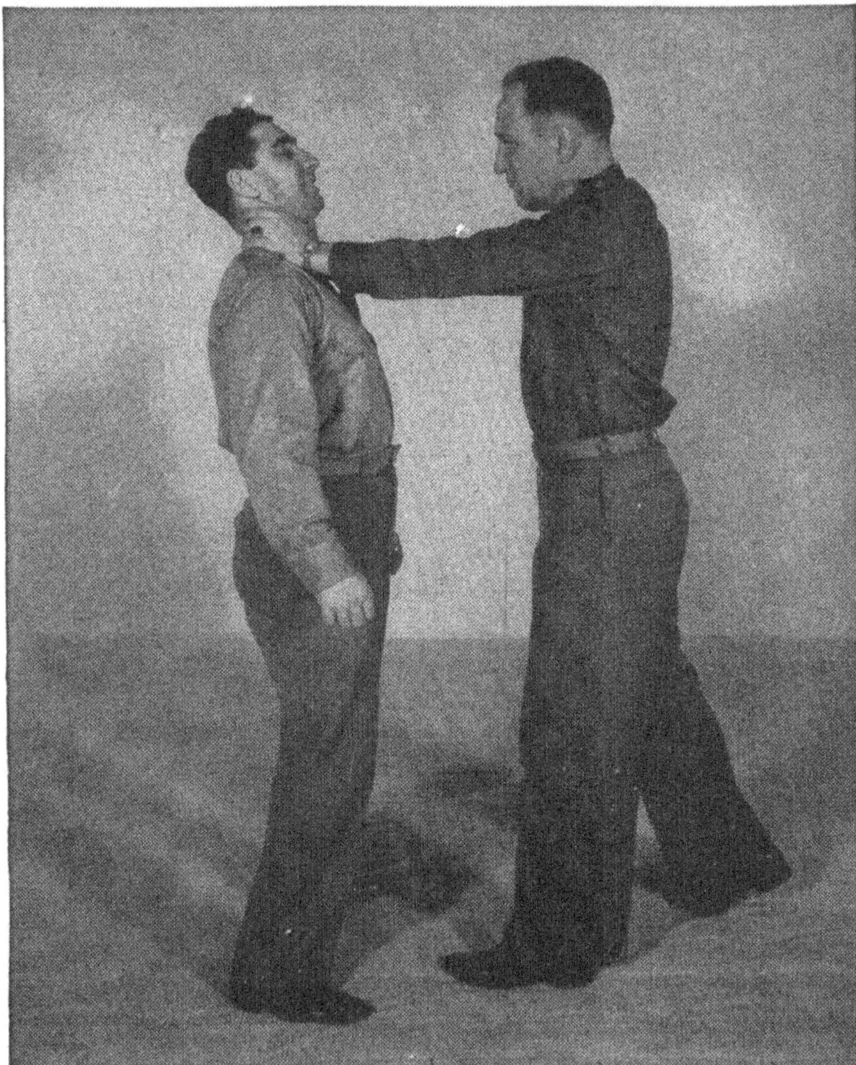

FIGURE 22 ①.

opponent grasps you around the throat from the front with both hands (fig. 22 ①). Reach up with your right hand, grasping the base of his thumb with your four fingers and the back of his hand with your right thumb, at the same time as you bend slightly to the right (fig. 22 ②). This movement

FIGURE 22 ②.

will loosen his grip on your throat and make it a simple matter for you to remove his right hand. Immediately reinforce the right-hand grip with a similar grip with your left hand, crossing your thumbs on the back of his hand. By applying pressure on the back of the hand toward the opponent's wrist, your opponent will be forced to the ground (fig. 22 ③). If

FIGURE 22 ③.

this entire action is made swiftly, your opponent will find himself on the ground a split second after he grasps you by the throat.

■ 28. **Second Defense Against Two-hand Front Choke.**—Your opponent again grasps you around the throat from the front with both hands. This time, instead of throwing him,

Figure 23 ①.

you wish to take him a prisoner. Your initial action is to reach under his left arm with your right forearm, placing your hand on top of his right hand. Your fingers grasp the palm of his right hand around the little finger edge, and your thumb is on the back (fig. 23 ①). Now turn to your right, pulling your opponent's right hand away from your throat

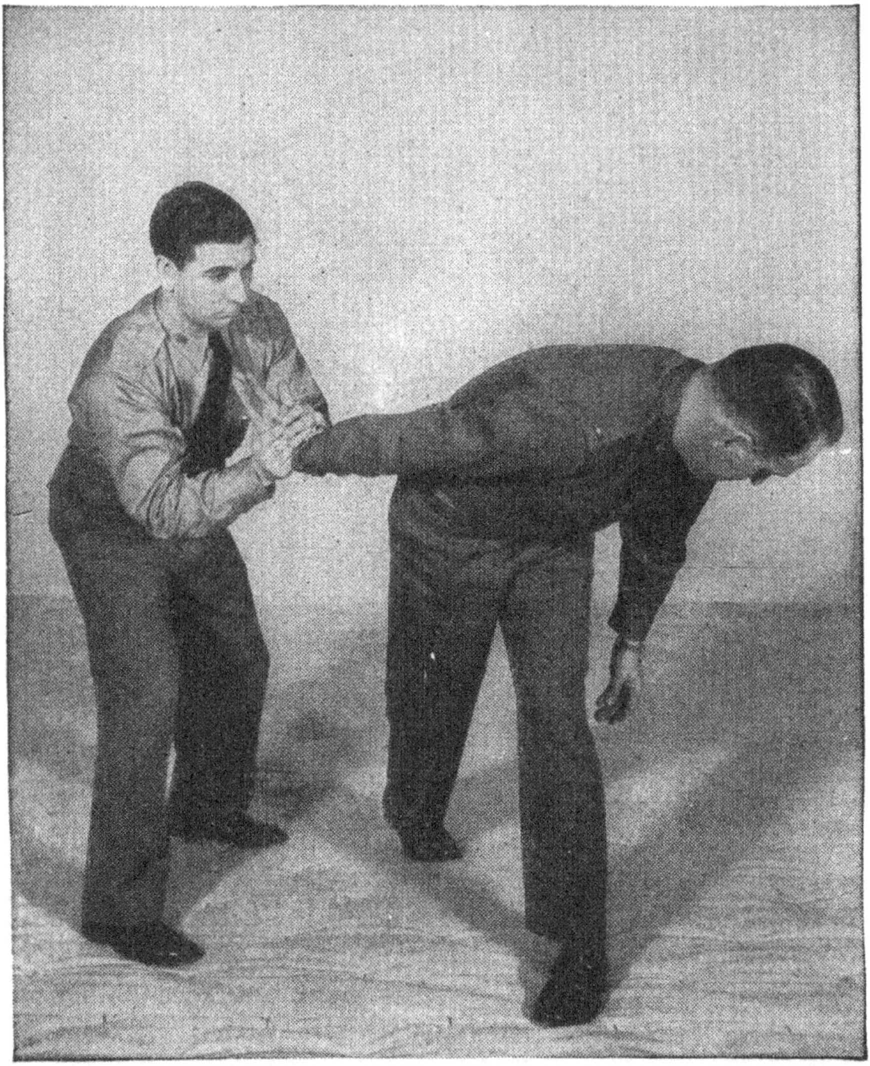

FIGURE 23 ②.

and turning his hand in the same direction as you are turning your body. This will cause him to turn his back to you. Now reinforce your original grip on his right hand with a similar grip with your left hand. Your defense will terminate as illustrated in figure 23 ②. In this position, he is your

FIGURE 24 ①.

prisoner and can be moved where you will by applying pressure toward his wrist.

■ 29. THIRD DEFENSE AGAINST TWO-HAND FRONT CHOKE.—Your opponent again grasps you around the throat from the front with two hands. Your defense this time will be such as to bring your opponent into a position where you can march him at a great distance without losing control. Your right

FIGURE 24 ②.

arm this time goes *over* his left arm and under his right one, the back of your hand resting underneath his right wrist. Your right hand forms a cup into which your fist will rest (fig. 24 ①). Strike upward with your left fist against your cupped right hand, thus breaking your opponent's right-hand grip on your throat and simultaneously loosening his left

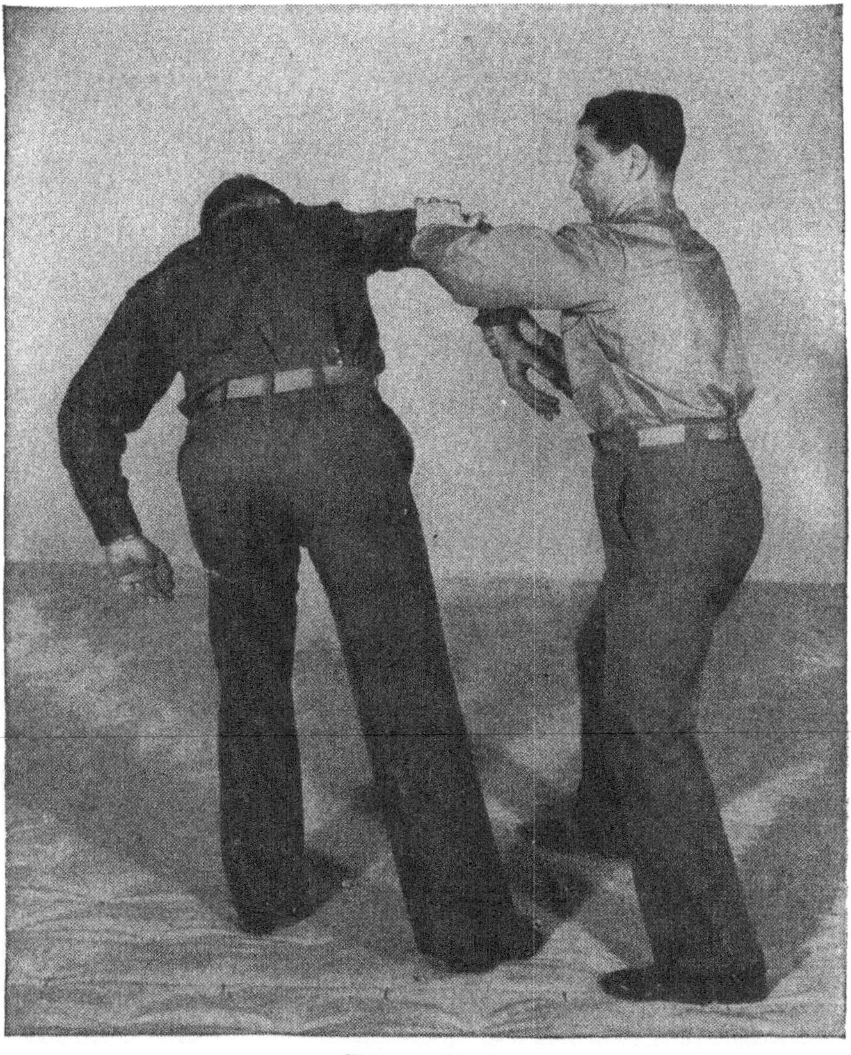

FIGURE 24 ③.

hand. As soon as the hold is broken, your right hand turns, grasping your opponent's right wrist with your thumb toward his body (fig. 24 ②). Now place your left hand behind your opponent's right elbow without actually grasping it and push with your left arm and pull with your right hand, causing your opponent to pivot, at the same time bending his right arm (fig. 24 ③). You now have sufficient control of his right

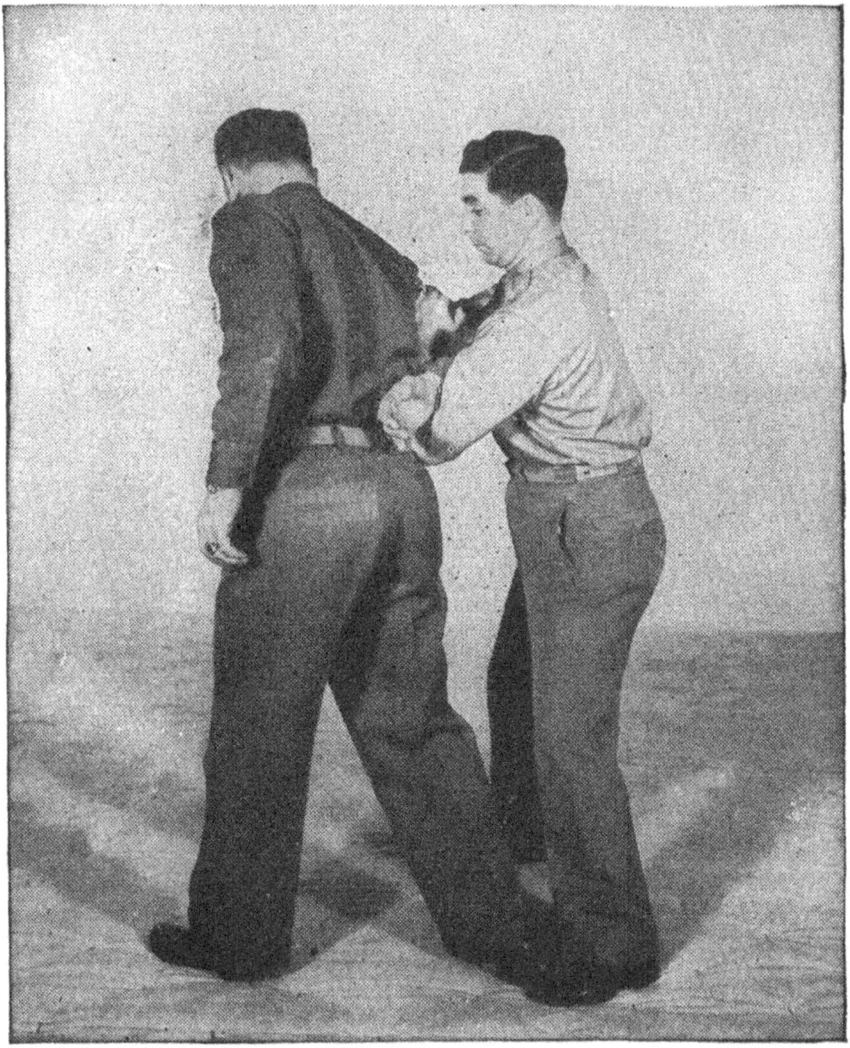

FIGURE 24 ④.

wrist to bend his arm at the elbow, bringing his right hand into the crook of your left elbow and slipping your left hand up his arm to a point near the shoulder which you can then grasp, terminating in the position illustrated in figure 24 ④. You can now apply extreme pressure, when necessary, forcing his right hand up his back. However, this procedure is not

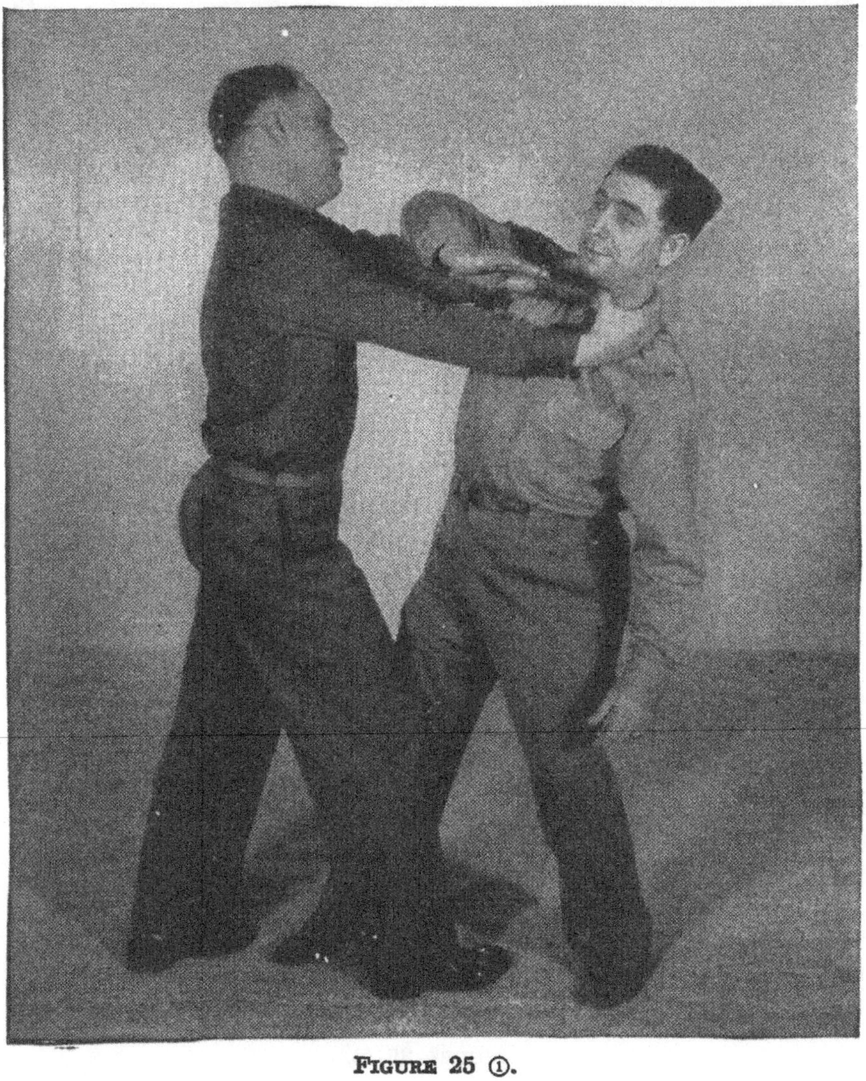

FIGURE 25 ①.

recommended except when meeting resistance. This is simply a means of marching a man where you wish him to go, utilizing but one arm and leaving the other one free for any necessary action.

■ 30. **Defense Against Two-hand Front Choke by a Tall Man.**—This will describe the proper defense when you are being choked by a very tall man, against whom some of the

Figure 25 ②.

previous defenses might not work. Step forward with your right foot, crossing your right arm with some force *over* both your opponent's arms, with the palm of your hand toward the ground (fig. 25 ①). By turning your body to the right, using the power of your midsection and shoulders, you will bring the little finger edge of your hand (not the little finger itself) against the right side of your opponent's neck just below the jawbone (fig. 25 ②). This is a knock-out blow.

FIGURE 26 ①.

■ 31. DEFENSE AGAINST TWO-HAND FRONT CHOKE BY A SHORT MAN.—You reach forward and *inside* your oponent's arms, grasping either his hair, or if he is bald, grapsing his ears (fig. 26 ①). You then pull downward on either hair or ears at the same time raising your knee. This will bring your opponent's face downward with great force as your knee

FIGURE 26 ②.

travels upward with an equal amount of power, meeting somewhere in between (fig. 26 ②). Since these two objects are approaching each other, the force is multiplied. Since your knee or thigh is definitely more capable of taking punishment than your opponent's face, the results are plainly evident.

■ 32. FIRST DEFENSE AGAINST TWO-HAND FRONT CHOKE AGAINST THE WALL.—Ordinarily, an individual who attempts to choke

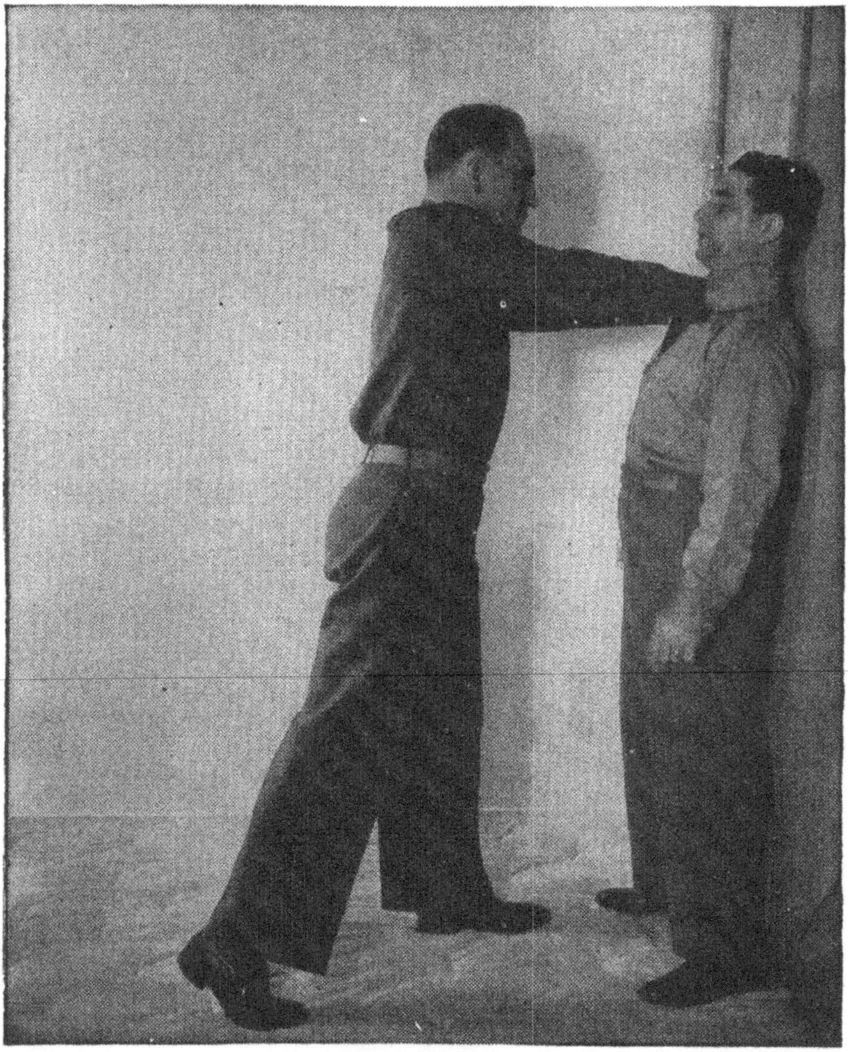

FIGURE 27 ①.

you in this fashion extends his arms, squeezes with the fingers, and pushes you against the wall (fig. 27 ①). Your immediate action is to bring the heel of your hands, one on either side of his elbows, applying pressure inward (toward each other) and away from you (fig. 27 ②). The reaction

FIGURE 27 ②.

is such as to prevent your opponent from using the power of his fingers, and he will find that he cannot choke you, try as he will.

■ 33. SECOND DEFENSE AGAINST TWO-HAND FRONT CHOKE AGAINST THE WALL.—You reach *under* your opponent's right arm and over his left one, placing your fingers on the outside of his

FIGURE 28 ①.

left elbow. You also place the palm of your right hand over the same elbow if necessary, over your own fingers (fig. 28 ①). By applying sharp pressure from the body on this elbow, you will force your opponent to your left, and, in most cases, crash his head against the wall (fig. 28 ②).

■ 34. THIRD DEFENSE AGAINST TWO-HAND FRONT CHOKE AGAINST

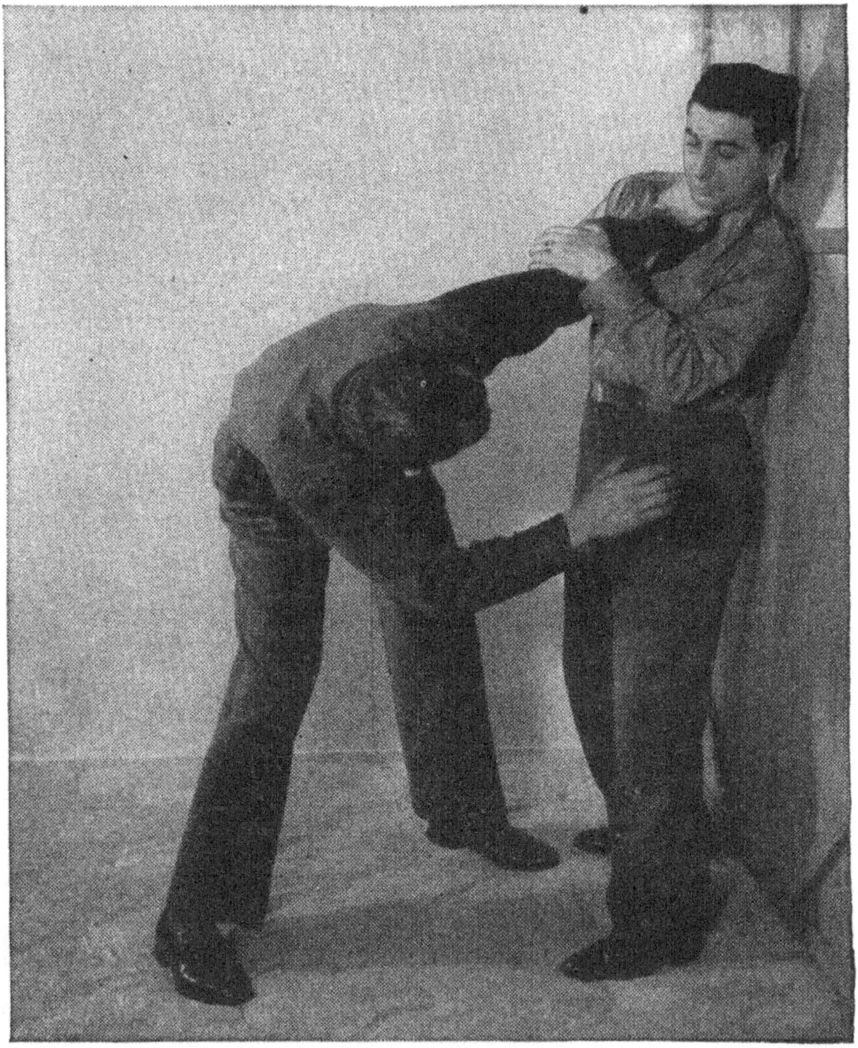

FIGURE 28 ③.

THE WALL.—If your opponent should grasp you in this manner and bend his arms, he will thus render the first two defenses useless. Your defense for such an attack will be to raise your left hand, the back of the hand toward the ground, and strike with the power of the body and shoulders, aiming the blow at the soft spot between the hip bone and

FIGURE 29.

the floating ribs. Contact is made with the little finger edge of the hand (fig. 29). This blow can be delivered by either hand on either side, depending on whether you are right-handed or left-handed. It will cause your opponent to release his hold, and, in most cases, also deprive him of his wind for a good space of time.

■ 35. DEFENSE AGAINST ONE-ARM STRANGLE FROM REAR.—Your

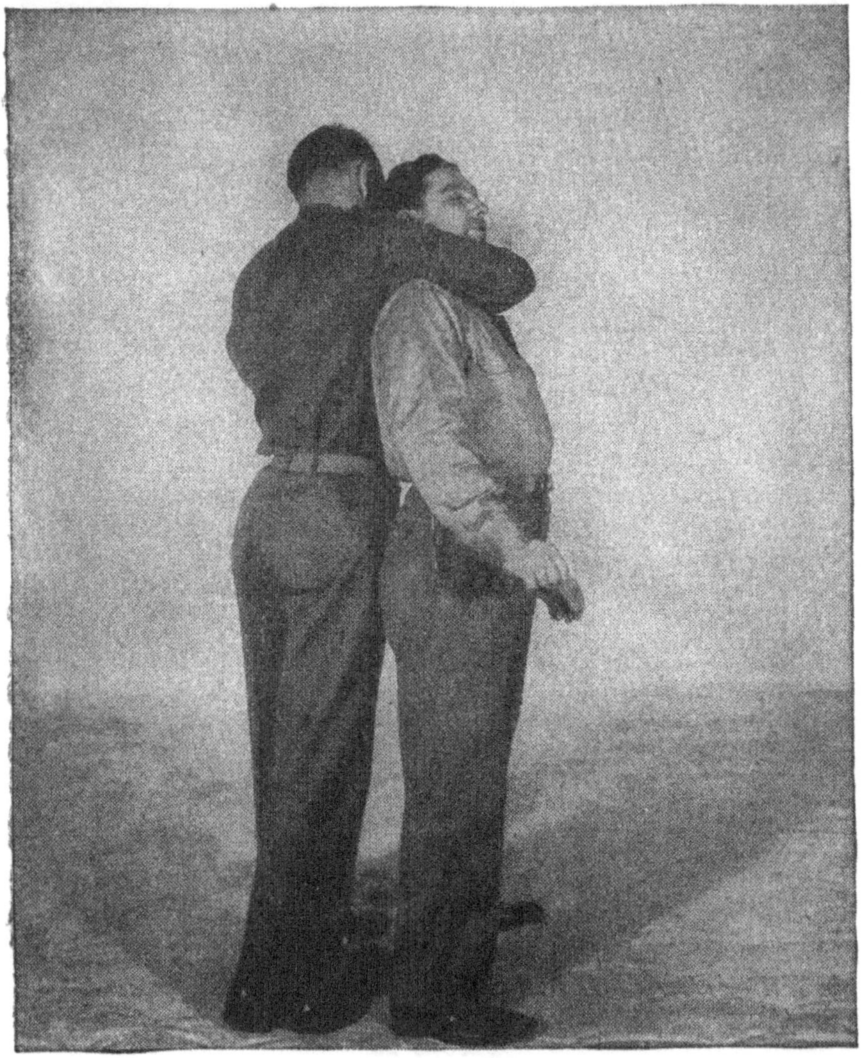

FIGURE 30 ①.

opponent attacks as illustrated in figure 30 ①. Your initial action is to place your chin in the crook of your opponent's elbow so that he cannot choke you, at the same time grasping the back of his arm just above the elbow with your right hand and placing your right foot just on the outside of his right foot (fig. 30 ②). By striking backward with your hips against

FIGURE 30 ②.

his midsection, retaining your hold on his upper arm with your right hand and bending from the waist swiftly, you will catapult your opponent over your head and to the ground (fig. 30 ③).

■ 36. **Defense Against Two-Hand Choke From Rear.**—Your opponent attacks in the manner illustrated in figure 31 ①. Your initial action is to reach up with your right hand, grasp-

Figure 30 ③.

ing the base of his right thumb and placing your own right thumb on the back of his hand close to the wrist, at the same time bending to the right (fig. 31 ②). You now balance on your right foot, pivoting on the sole of that foot to the right, using your left leg for momentum. As soon as you have executed half a right turn, you will reinforce your right-hand grip on your opponent's right hand with a grip with your

FIGURE 31 ①.

left hand, twisting strongly to your left (fig. 31 ③). This will result in your opponent either executing a somersault, usually landing on his head, or having his wrist dislocated or broken.

■ 37. DEFENSE AGAINST ONE-HAND FRONT CHOKE.—Your opponent attacks by grasping you by the throat with his right hand and attempting to choke you as illustrated in figure

FIGURE 31 ②.

32 ①. You step backward with your right foot, at the same time grasping his right wrist with your right hand and *holding it close to your throat*. Place your left foot between and in front of his legs (fig. 32 ②). Bring your left arm over

your opponent's right elbow without touching it with your hand; retaining your grasp on your opponent's right wrist with your right hand, you bring your armpit on top of his

FIGURE 31 ⑧.

right elbow, and turning to the right, reinforce the grasp on his wrist with a similar grasp with your left hand. Now bring pressure to bear by lifting upward on the captured right wrist and pushing downward with your armpit on his elbow. In this position, you can either force him to submit, or to suffer a broken or dislocated elbow.

FIGURE 32 ①.

FIGURE 32 ②.

95

FIGURE 32 ③.

SECTION VI

DEFENSES AGAINST KICKS

■ 38. **FIRST DEFENSE AGAINST KICK WITH RIGHT FOOT.**—When you are just within reach of your opponent's foot, and he

FIGURE 33 ①.

suddenly reaches out, attempting to kick you in the groin, you execute a side movement of your body without moving your feet. This is done by bending the left knee, straightening the right knee, and twisting the hips to the right as illustrated in figure 33 ①. This causes your opponent's foot

FIGURE 33 ②.

to miss its mark. You make no attempt to stop the force of the kick, but rather place your hands under his foot and ankle and continue his movement by lifting upward (fig. 33 ②). His momentum plus even a slight amount of additional help by your hands will lift your opponent into the air, dropping him on his back (fig. 33 ③).

■ 39. SECOND DEFENSE AGAINST KICK WITH RIGHT FOOT.—When you are close to your opponent, and he attempts to knee you

FIGURE 33 ③.

or use a short kick, you will turn your body to the right slightly, raising your left knee to your right front as a means of protection and bring the outside edge of the sole of your shoe to catch the blow (fig. 34 ①). When your opponent's shin makes contact with the hard edge of your shoe, the results will be painful and effective (fig 34 ②).

FIGURE 34 ①.

■ 40. DEFENSE AGAINST KICK WITH LEFT FOOT.—Since you can never be sure with which foot your opponent intends to kick, the initial side movement of your body must always be in the same direction. Therefore, your initial movement in this case will be identical to that in figure 33 ①. Figure 35 ① illustrates the movement. The only difference is that in this case, immediately upon executing the side movement of

FIGURE 34 ②.

the body, you will reach down with your right hand and catch your opponent's ankle, at the same time stepping forward with your right foot (fig. 35 ②). From this position, you will balance on your right foot, and lifting your left foot, strike forcibly with the calf against the back of your opponent's knee, at the same time striking him under the chin with the heel of your left hand (fig. 35 ③). The results are self-evident.

FIGURE 35 ①.

FIGURE 35 ②.

FIGURE 35 ③.

Section VII

TAKING PRISONERS

■ 41. Handling a Hand-Squeezer.—Figure 36 ① illustrates a powerful man squeezing with such force as to cause your

Figure 36 ①.

knees to buckle. Your initial movement is to bend the knees slightly, pretending to give way, raise your opponent's right hand to the right just far enough to enable you to take a long step to his right rear, passing your head under his armpit (fig. 36 ②). As soon as your head is on the rear side of

FIGURE 36 ②.

his arm, you will bring your left arm over his right one, placing your armpit on his elbow and placing the palm of your hand against the back of his right hand. All this time you have retained the original hand clasp with your right hand.

You can now subdue your opponent by lifting up on the captured hand and pressing downward with the armpit against his elbow (fig. 36 ③).

FIGURE 36 ③.

■ 42. EJECTING A TROUBLESOME VISITOR.—It is, at times, necessary to eject an individual from an establishment without causing a great deal of commotion. This is usually done in cases of an obstreperous drunk. A drunk is always willing to shake hands. Therefore, you will come toward him and grasp his right hand with your right one (fig. 37 ①). If he doesn't want to grasp hands, grasp his wrist from the inside.

FIGURE 37 ①.

You will then lift the captured right arm just high enough to permit you to pass your left arm under it, at the same time stepping to your opponent's right with your left foot (fig. 37 ②). Now grasp his left lapel or his left collar and by pulling downward and backward on the captured right hand or wrist, you will cause pressure to be brought on his ulna nerve by contact with your wrist or forearm (fig. 37 ③). In

FIGURE 37 ②.

this position, it is a simple matter to *walk* even the strongest of men without creating a disturbance.

■ 43. A SIMPLE COME-ALONG.—A simple method of taking a prisoner is to grasp his right wrist with your right hand (fig.

38 ①), taking a step with your left foot to his right side, at the same time pulling the captured right arm toward you. This time, you will reach your left upper arm *over* his upper

FIGURE 37 ③.

arm (fig. 38 ②), and bring your forearm or sharp wrist bone *under* the back of his upper arm, locking your hand on your chest. Keeping his thumb uppermost and applying pressure downward on the captured right wrist will expose the ulna nerve to contact with your left wrist or forearm, bringing your opponent to his toes and making it simple to take him along (fig. 38 ③).

FIGURE 38 ①.

■ 44. ALTERNATE FOR SIMPLE COME-ALONG WHEN MEETING RESISTANCE.—When attempting a simple come-along, occasionally your opponent will frustrate your attempt by bending his elbow as illustrated in figure 39 ①. You will then simply utilize the energy expended by him in bending his arm by continuing its upward course, retaining your hold on

FIGURE 38 ②.

his right wrist with your right hand by sliding your hand around. Then bring the edge of your left forearm *hard* against the crook of his right elbow, causing it to bend further, and lock your left hand on your right wrist or forearm (fig. 39 ②). You will then bring your elbows as close as possible to your body, making sure that one of your elbows is on either side

FIGURE 38 ③.

of the captured right arm. It is now a simple matter to bring your opponent to the ground simply by bending from the hips (fig. 39 ③).

■ 45. TAKING AN APPROACHING MAN PRISONER.—This hold is very effective when you wish to take an approaching individual prisoner with such rapidity that he will not have time to

FIGURE 39 ①.

utilize his weapons. You will pretend to pass on his left side, apparently ignoring him, but as you come opposite, raise your left hand and grasp his biceps just above the elbow with your thumb on the outside of his arm and the fingers on the inside, at the same time grasping the two smallest fingers of his left hand with the four fingers of your right hand (fig. 40 ①).

FIGURE 39 ②.

Immediately pivot to the left, retaining your holds, placing your thumb between the middle and the next smallest fingers of his left hand (fig. 40 ②). Bringing his arm upward, causing it to bend at the elbow and wrist, you will anchor his elbow between your right arm and body, at the same time applying pressure on the last two fingers of his left hand. Pressure

FIGURE 39 ③.

will be brought to bear toward the wrist and elbow bringing him to his toes and making it possible to walk him anywhere you please. (fig. 40 ③). If he should attempt to strike you or reach for a weapon with his right hand, you can stop any action by applying a greater degree of pressure on the captured left wrist.

■ 46. FINGERS COME-ALONG.—This is a come-along that is

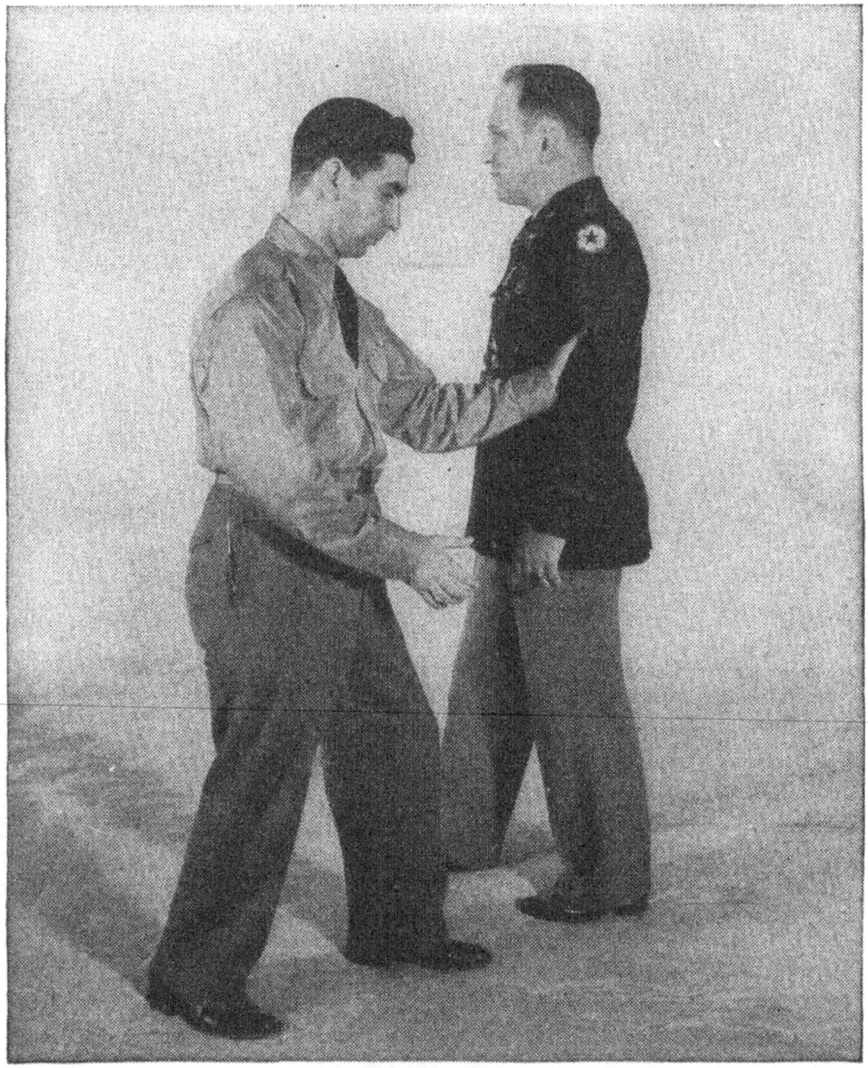

FIGURE 40 ①.

very easy to get, but it cannot be utilized for any great distance. You reach for your opponent's right hand with your left one, grasping either two or three of his fingers (fig. 41 ①).

Grasping the fingers tightly, turn the palm of your opponent's hand upward and bend his fingers backward toward his wrist, maintaining control with the power of your elbow

FIGURE 40 ②.

close to your body (fig. 41 ②). By applying a greater degree of pressure, your opponent will be forced to his toes to prevent his fingers from being broken and you will be enabled to march him where you see fit. If he attempts to turn, simply apply a greater amount of pressure.

■ 47. LOCK COME-ALONG.—This come-along can be used for

FIGURE 40 ③.

walking a dangerous individual on long distances. As you come opposite your opponent, place your right wrist against his left wrist (without attempting to grasp it) and your left hand will grasp his upper arm just above the elbow, your fingers to the rear (fig. 42 ①). Now push to the rear with your right forearm and pull toward you with your left hand,

FIGURE 41 ①.

which will cause your opponent to bend at the waist and also cause his arm to bend upward and to the rear (fig. 42 ②). As your opponent's left arm comes to a right angle up his back, you will be in position to push your right arm under his left forearm and over his upper arm, grasping his shoulder, terminating in the position illustrated in figure 42 ③.

FIGURE 41 ②.

■ 48. WRIST REVERSE COME-ALONG.—Reach across with your *left* hand and grasp your opponent's *left* hand with your fingers around the palm of the little finger edge and your thumb on the back of his knuckles (fig. 43 ①). Now bring your left elbow close to your low abdomen and turn your opponent's hand to your left (fig. 43 ②). Then reinforce

FIGURE 41 ③.

your original hold with a similar hold with your right hand, terminating in a position where the fingers of both your hands will be grasped on the palm of his left hand and your thumbs will be crossed on the back of his knuckles. Pressure is applied toward his wrist (fig. 43 ③).

All the holds in this section may be used on either side.

FIGURE 42 ①.

FIGURE 42 ②

FIGURE 42 ③.

FIGURE 43 ①.

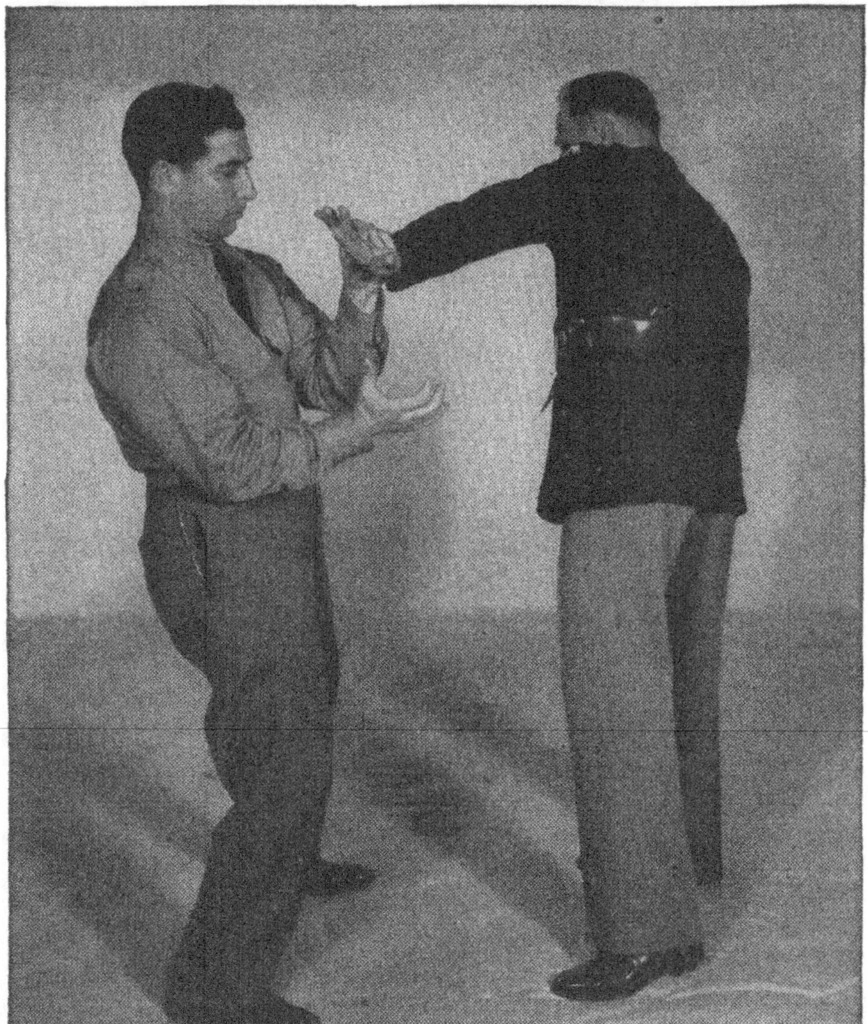

FIGURE 43 ②.

FIGURE 43 ③.

SECTION VIII

DEFENSES AGAINST KNIFE AND SWORD

■ 49. DEFENSE AGAINST DOWNWARD STROKE OF KNIFE.—Your

FIGURE 44 ①.

opponent attacks in the manner illustrated in figure 44 ①. Only an angry man or an individual who has had no training in the use of the knife would use this mode of attack, since the blade is very likely to be deflected by the breast bone or ribs. You ward off the blow by presenting your right forearm against the striking arm of your opponent and strike him in

FIGURE 44 ②.

the crook of the right elbow with the thumb side of your left forearm or wrist (fig. 44 ②). This will cause his arm to bend. Then grasp your right wrist with your left hand, bring your elbows close to your body, and bending swiftly from the waist, bring extreme pressure to bear on your opponent's knife arm, causing him to fall backward and lose his weapon.

FIGURE 44 ③.

■ 50. DEFENSE AGAINST UPWARD STROKE OF KNIFE.—The individual who uses the knife in the manner illustrated in figure 45 ① is extremely dangerous, since he strikes upward with a short snappy movement, aiming at the soft parts of the abdomen. If he strikes home, he is likely to impose a fatal wound. You block the stroke by bringing the little finger edge

FIGURE 45 ①.

of your *left arm* against your opponent's wrist or forearm. As an added precaution, a simultaneous movement will be the side twist of the body as described in previous sections. Right leg is straightened, left knee is bent, and your body turns to the right. Immediately the blow is stopped, the right hand grasps your attacker's right hand with your thumb on the

FIGURE 45 ②.

back of his knuckles and your fingers around the palm side of his hand. Contact is retained with your left arm against his left forearm (fig. 45 ②). You now reinforce your right-hand hold with a similar hold with your left hand which will result in your thumbs crossing on the back of his knuckles and the fingers of both your hands meeting in the center of the palm

FIGURE 45 ③.

of his hand. You will now twist his wrist to your left and bend his hand at an angle of 45° toward his forearm (fig. 45 ③).

■ 51. **Defense Against Back-hand Slash With Knife.**—A very common attack with the knife is the one illustrated in figure 46 ①. Two simultaneous actions are necessary in your defense.

Figure 46 ①.

The knees are bent and the body lowered without ducking your head; at the same time, your right arm is raised and the forearm or wrist presented against the forearm or wrist of the striking arm of your attacker (fig. 46 ①). Immediately the blow is stopped, your left hand reaches over your head, grasps your opponent's knife hand with your fingers around the palm of his hand, and your thumb in the center of the back of his

FIGURE 46 ②.

knuckles. Pressure is continued with your right wrist against his right wrist. You now start to twist the captured knife hand to your left (fig. 46 ②). You then reinforce your left-hand hold with a similar hold with your right hand, thumbs in the center of the knuckles and fingers around the palm or hilt of your opponent's knife hand. A twist to your left or pressure at an angle of 45° bending your opponent's hand

FIGURE 46 ③.

against his wrist will cause him to lose his weapon, and, in many cases, suffer a broken or dislocated wrist.

■ 52. **Defense Against Wary Approach With Knife.**—Your opponent attacks in the manner illustrated in figure 47 ①. His left foot is extended and his left hand is extended to ward off

Figure 47 ①.

any of your possible blows. His knife hand is held close to his hip, ready to strike when an opening occurs. This is an extremely dangerous antagonist. He is well-prepared and well-trained and your defense must be perfect. If it is possible to use anything as a weapon, a chair, table, tobacco, dust, or any object at all with which you can strike him, throw in his eyes, or wrap around his arm, it is advisable to use it. However, if you are not in a position where you have such a defense, you must use the following attack. Immedi-

FIGURE 47 (2).

ately your opponent comes within arm's length of your nearest leg, you will spring from the ground, launching your body at him feet first, and turning to your left. You left instep will hook around your opponent's left ankle. Your right foot will kick him strongly on the inside of the knee or thigh, while you break the force of your fall to the ground with your hand or arm (fig. 47 ②). This scissoring motion done rapidly and taking your opponent by surprise will drop him on his back. The best he can possibly do with his knife is cut at your feet

FIGURE 47 ③.

or lower leg. Immediately you both strike the ground, you will raise your right foot and bring your heel viciously into his groin or midsection (fig. 47 ③).

■ 53. LUNGE WITH THE SWORD.—An individual trained in the use of the foil or épée will approach you in the manner illustrated in figure 48 ①. While swords are not used to any great extent in modern warfare, the opportunity might present itself when you will have the need for this defense. As your

FIGURE 48 ①.

opponent lunges with the sword, with the point aimed at your midsection, you execute the sideward movement of the body, previously described, to the right, at the same time tapping the blade of the attacking weapon smartly with either your forearm or wrist (the weapon should not be grasped with the hand (fig. 48 ②). Immediately you have tapped the sword aside, you will take a short step with your left foot and immediately raise your right leg, *kicking* your opponent behind the

FIGURE 48 ②.

knee joint forcibly, at the same time that you bring the heel of your right hand against your opponent's chin (fig. 48 ③).

■ 54. DEFENSE AGAINST STRAIGHT SABER CUT.—The average saberman will attempt the downward cut illustrated in figure

49 ①. Your immediate action is to step as close to your opponent as possible, bringing your left forearm against his striking arm (fig. 49 ②). Immediately you have broken the force

FIGURE 48 ③.

of the blow, raise your right knee and your right hand simultaneously. The fingers of your right hand, palm uppermost, will be brought with great force into your opponent's solar plexus while your knee is simultaneously brought into his groin (fig. 49 ③).

■ 55. **Defense Against Cross-cut With Saber.**—It is unusual for a trained individual to attempt a cut illustrated in figure 50 ①, but if the attempt is made, you will step with your

Figure 49 ①.

right leg and present your right forearm against your opponent's striking arm. You will make no attempt to stop the blow directly, but will continue its momentum in the direction in which it is moving, stepping to your opponent's right rear with your left leg, and reaching over his left shoulder with your left hand (fig. 50 ②). You will now be in a position behind your opponent, grasping his wrist with the palm turned away from your body. Grasp his right lapel with your

FIGURE 49 ②.

left hand, crossing over the front of your opponent's throat. Simultaneously, raise your right leg and placing the sole of your foot against the back of your opponent's right knee joint push strongly. This action will topple your opponent backward and place you in a position to break his right elbow over your thigh or hip (fig. 50 ③).

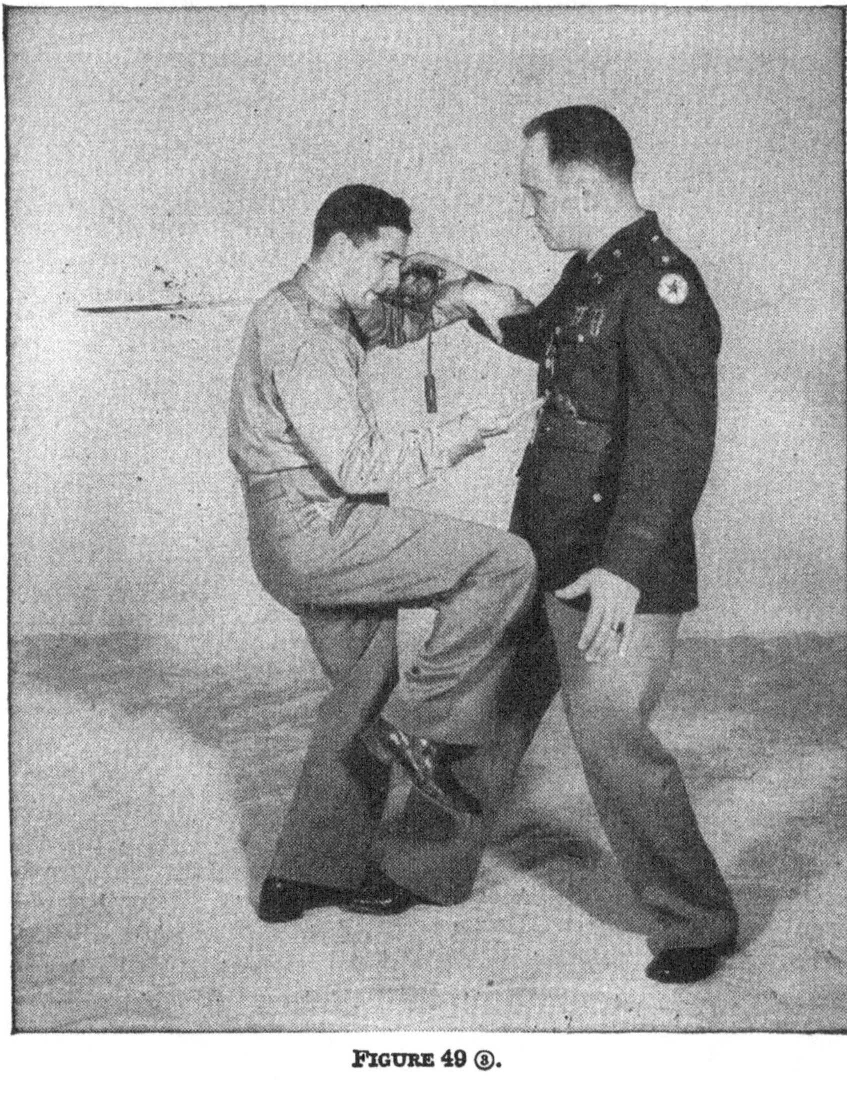

FIGURE 49 ③.

FIGURE 50 ①.

FIGURE 50 ②.

FIGURE 50 ③.

Section IX

DEFENSES AGAINST BLOWS WITH CLUB, AND TECHNIQUE OF CLUB

■ 56. **Defense Against Downward Blow With Club.**—As

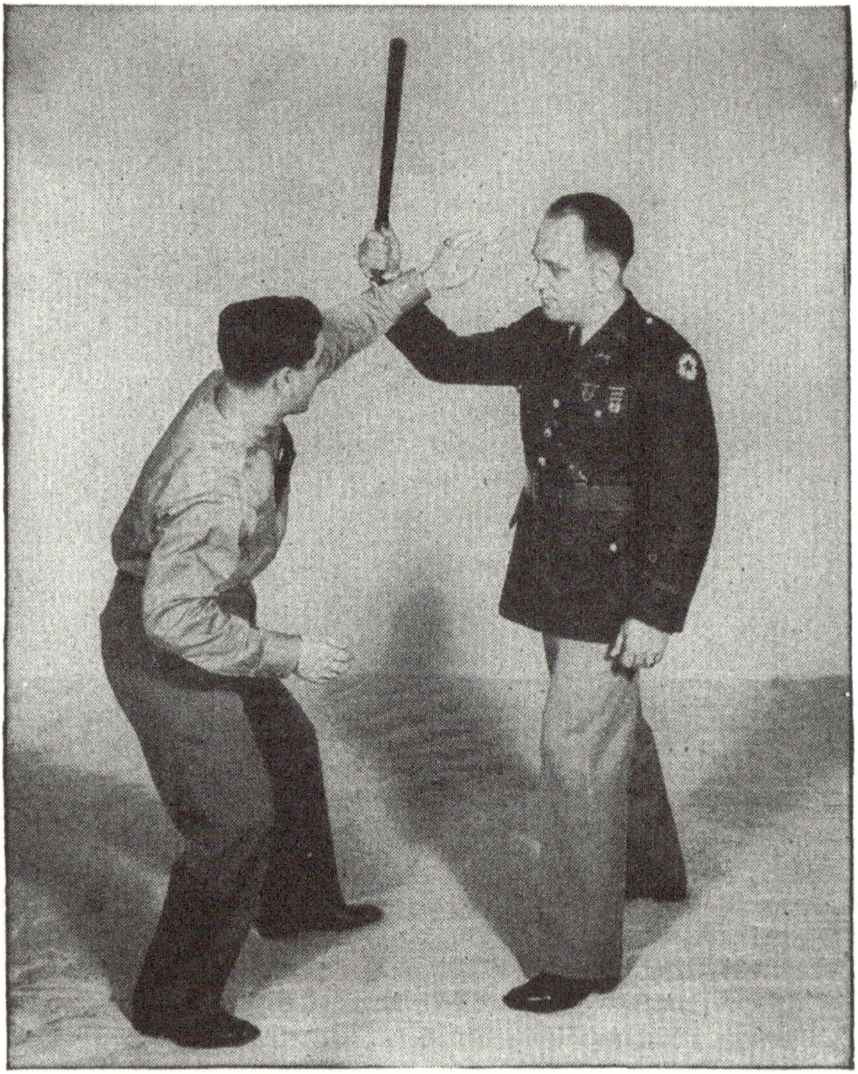

Figure 51 ①.

your opponent strikes a downward blow with the club, you will present your left forearm against his right forearm in the manner illustrated in figure 51 ①. You will make no attempt to stop the blow directly, but rather deflect it to your left so as to enable you to wrap your left arm around his

FIGURE 51 ②.

right one, bringing your wrist under his right elbow (fig. 51 ②). Your further action is to place your right hand on your opponent's shoulder or upper arm and lock your left hand on your own forearm (fig. 51 ③). You are now in a position to break your opponent's arm simply by applying pressure. A small degree of pressure will cause him to drop his weapon.

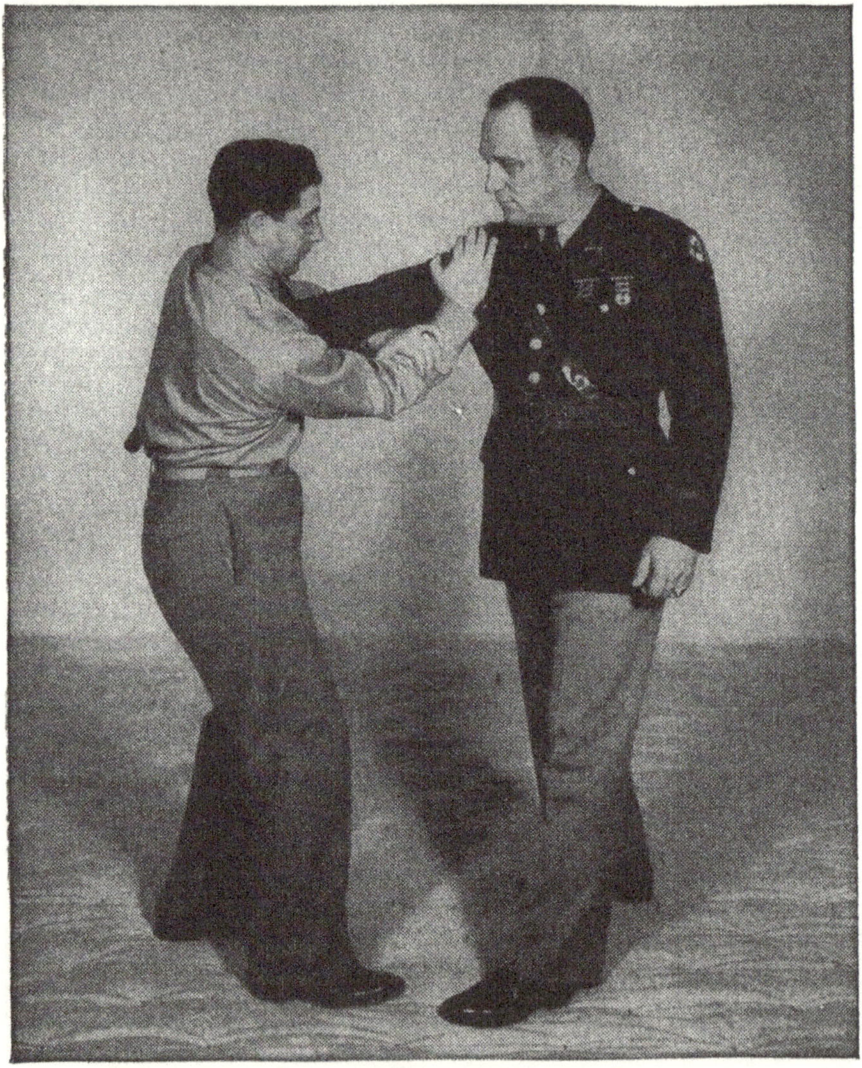

FIGURE 51 ③.

■ 57. DEFENSE AGAINST SIDE STROKE WITH CLUB.—The more experienced individual will strike a slanting sideward blow at the side of the head as illustrated in figure 52 ①. You will make no attempt to stop the blow, but will lower your head out of range by bending the knees, at the same time reaching upward with your left hand or arm (without grasping your opponent's club arm) and striking your opponent's

FIGURE 52 ①.

forearm, continuing its momentum over your head (fig. 52 ②). This will turn your opponent completely off balance. You will now take a long step with your right leg to your opponent's right, at the same time twisting your body to your left and raising your right arm shoulder-high with the back of your body and shoulders. Your elbow will make contact in the soft spot on your opponent's side between the hip bone

FIGURE 52 ②.

and the short ribs (fig. 52 ③). A man struck in this manner will drop as though shot.

■ 58. HOW TO HOLD CLUB.—The club, when it is carried, should be used only with the left hand. The thong should be of a length suited to the hand of the individual who is to use it. In wrapping the thong around the hand or arm, the following procedure should be followed. The thumb is first hooked

FIGURE 52 ③.

through the loop of the thong (fig. 53 ①). The thong is then brought over the *back* of the hand (fig. 53 ②), and the handle of the club brought up from the little finger edge and then grasped by the hand with the grip illustrated in figure

53 ⑧. The club should not be used as a bludgeon except in dire necessity, but should be used as an extension of the arm. It is a much more effective weapon if it is used to jab

FIGURE 53 ①.

rather than to strike. Practice in using the club in this manner will render it very effective as a defense against attack by many types of weapons. It can then be used in parrying blows or turning aside thrusts in the same manner as the fencer uses the foil. When it is necessary to stop a charging opponent or to subdue a recalcitrant individual, a jab to the solar plexus is extremely efficient (fig. 54 ①). If your oppo-

FIGURE 53 ②.

nent is so close as to render the body jab impractical, the chin or throat jab is equally effective (fig. 54 ②).

■ 59. WHERE TO STRIKE IF NECESSARY.—If an opponent or opponents are moving in and you do not wish to damage them severely, their desire for combat can often be cooled by the blow to the wrist or hands (fig. 55 ①), or by the blow to the shin (fig. 55 ②). If it becomes necessary to put a dan-

FIGURE 53 ③.

gerous antagonist out of action, the blow to the side of the throat just under the jaw line (fig. 55 ③), is efficient without being dangerous. The backhand blow to the opposite side of the throat (fig. 55 ④), is equally effective.

■ 60. **Using Club as Come-Along.**—The most efficient use of the club is its use as an adjunct in taking a man prisoner without causing him bodily harm. The club is still in the

Figure 54 ①.

left hand. Your right hand grasps your opponent's right wrist, and lifting his arm slightly, the club is thrust under his arm (fig. 56 ①), over his shoulder with its end lodged just behind your opponent's neck (fig. 56 ②). Your hand grasping the hilt of the night stick should be just behind your opponent's upper arm with your thumb against his ulna nerve two inches above the elbow. Pressure is applied downward

FIGURE 54 ②.

and backward with your right hand on your opponent's right wrist. Your opponent will be forced to his toes by the pain. However, he will be in no danger of having his arm broken (fig. 56 ③).

■ 61. **Using Club as Handcuff.**—You have utilized the standard procedure of searching a prisoner, placing his hands against the wall and causing him to extend his legs as far to

Figure 55 ①.

the rear as possible. Since he is dangerous, you now wish to *secure* him in such a manner as to enable you to march him where you will. You will place your right foot just inside of your opponent's right foot in order to be in position to drop him to the ground if he becomes belligerent, and order him to bring his right wrist backward, at which time you will grasp it with your right hand (fig. 57 ①). Now place one loop of the thong of your club over his right wrist and bend

FIGURE 55 ②.

his right arm at right angles up his back (fig. 57 ②). Order him to place the top of his head against the wall and to bring his left arm backward. If he fails to obey this order, a slight push on his right arm up his back will change his mind.

When he does bring back the left arm, you will insert that hand through the thong also (fig. 57 ③). You will now slip the handle of the club as close to your opponent's body as pos-

FIGURE 55 ③.

sible with the long part of the club extending up the back almost to the neck. Grasping the hilt of the club (fig. 57 ④), you will be in position to march your prisoner where you will.

Figure 55 ④.

FIGURE 56 ①.

FIGURE 56 ②.

FIGURE 56 ③.

FIGURE 57 ①.

FIGURE 57 ③.

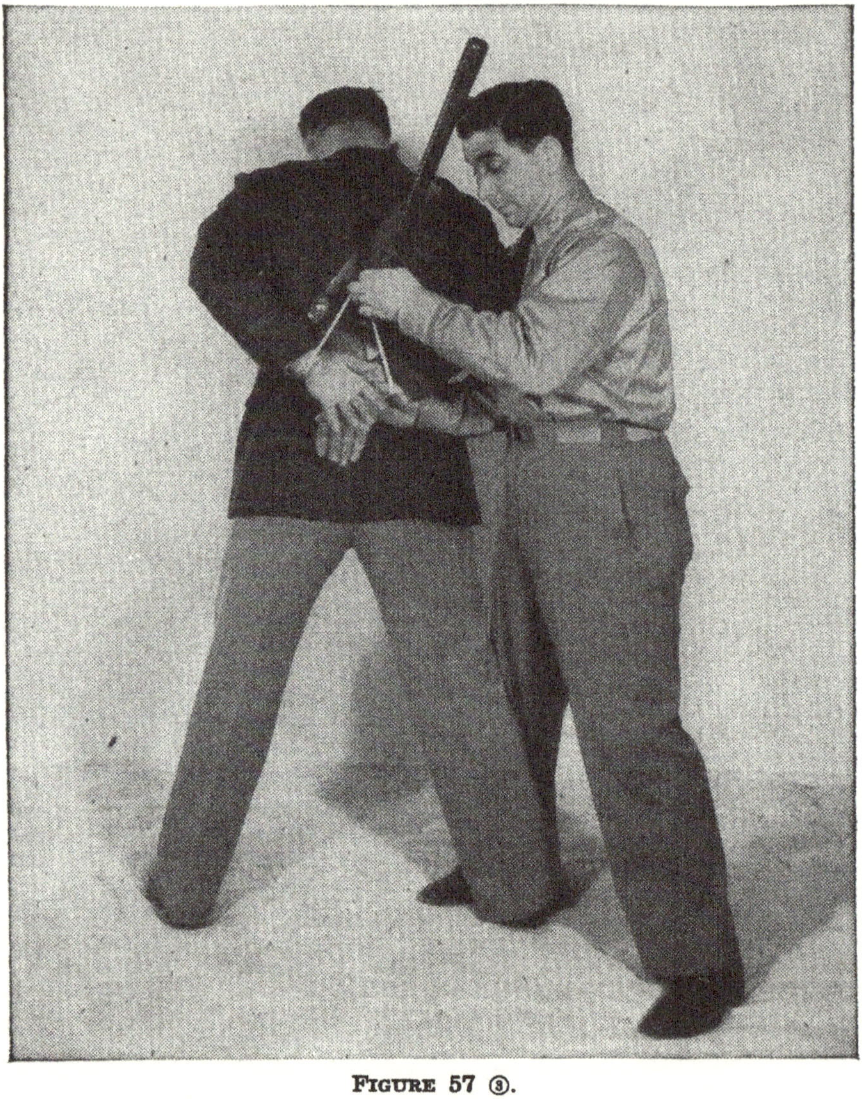

FIGURE 57 ③.

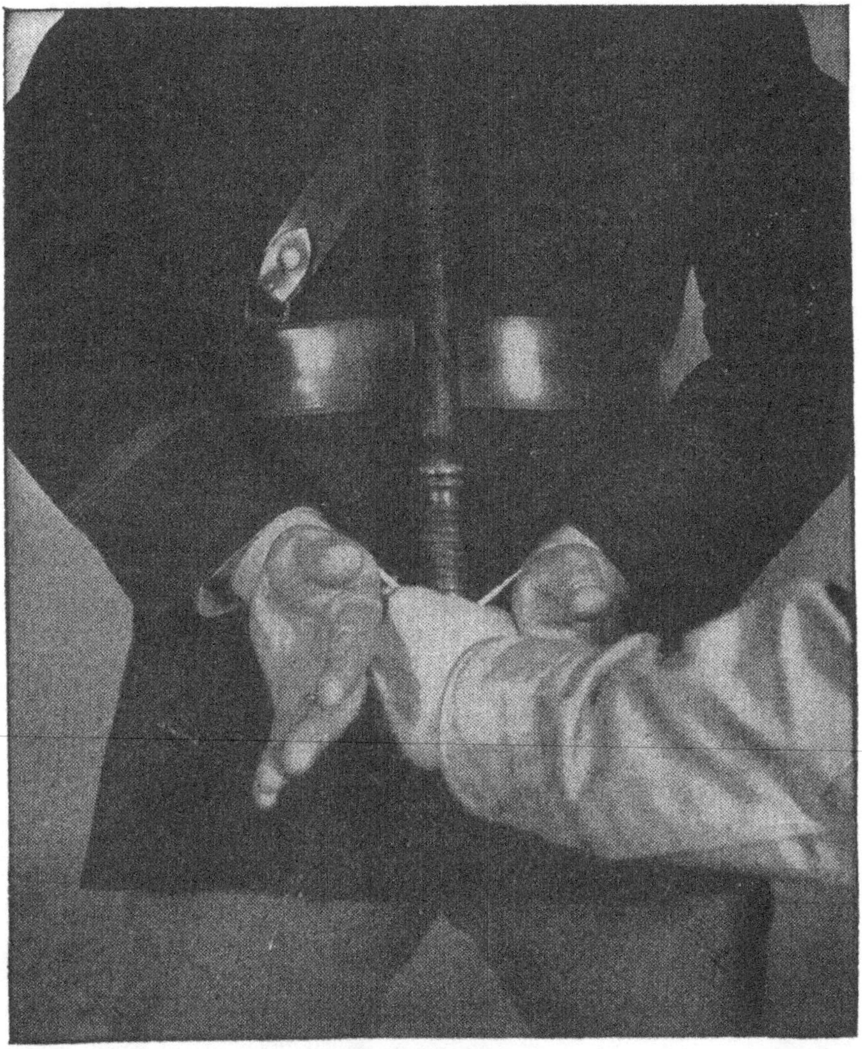

FIGURE 57 ④.

Section X

DEFENSES AGAINST PISTOL

■ 62. **Defense Against Gun in Front.**—Your opponent orders

Figure 58 ①.

you to raise your arms as illustrated in figure 58 ①. Note that while the hands are raised, the elbows are maintained as low as possible. You immediately use your side body pivot as previously described and strike your opponent's wrist with your *left* forearm (fig. 58 ②). You will immediately grasp the barrel of the gun, assuring yourself that your hand is not

FIGURE 58 ②.

near the muzzle, and at the same time push downward on your opponent's right wrist (fig. 58 ③). By maintaining pressure with your left hand, you will now have excellent fulcrum with your right hand to bend the gun backward, causing your opponent to release his grip (fig. 58 ④). If he should retain his grip, his index finger will be broken. From this position you

FIGURE 58 ③.

can bring the butt of the gun against your opponent's right temple.

■ 63. FIRST DEFENSE AGAINST GUN IN BACK (RIGHT HAND).—This defense should be used only when you are certain that the gun is in your opponent's right hand (fig. 59 ①). Also note that again the elbows are as close to the waist as possible, even though the hands are elevated. You again execute the

FIGURE 58 ④.

body side twist, at the same time bringing your elbow against your opponent's forearm (fig. 59 ②). Note that the feet have not moved. The right arm is then brought under your opponent's right forearm and placed on his elbow joint in such a manner that his forearm rests in the crook of your right elbow (fig. 59 ③). You will now place your left hand on top of your right and bend swiftly from the waist. This will bring

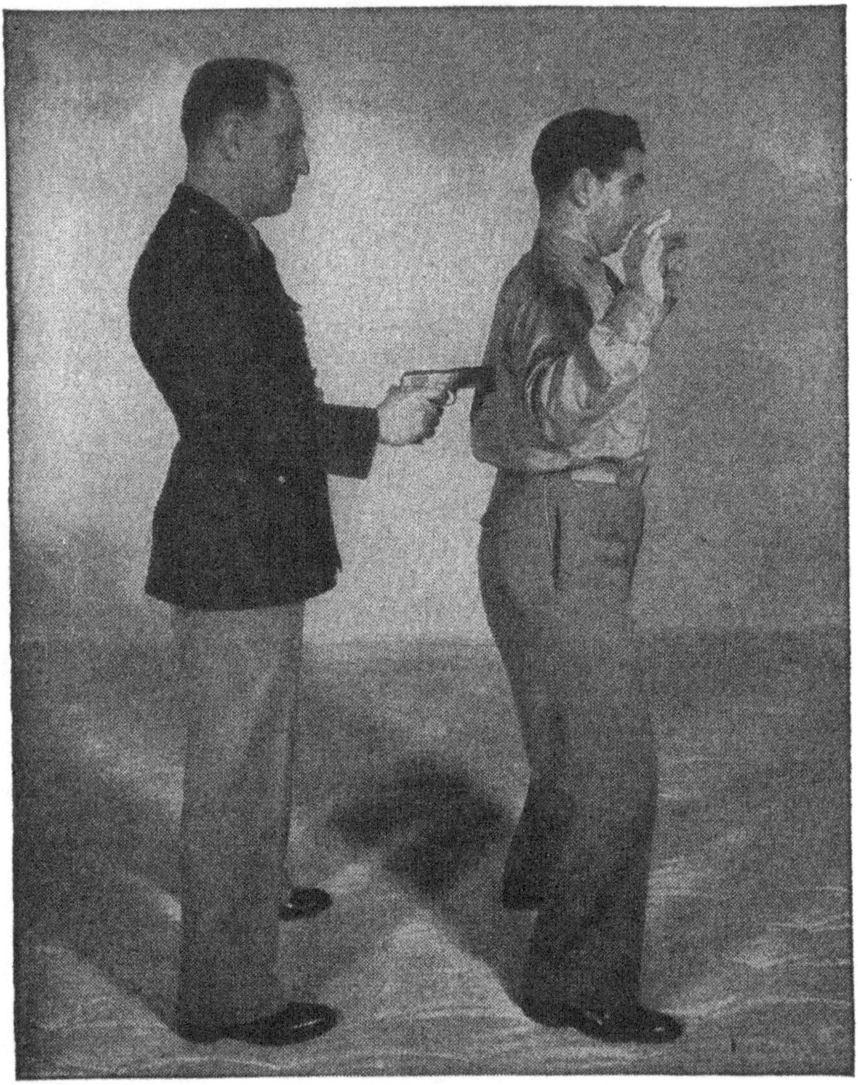

FIGURE 59 ①.

your opponent to the ground and force him to drop his weapon (fig. 59 ④).

■ 64. **Second Defense Against Gun in Back (Right Hand).**—In this defense also you must be sure that your opponent is holding the gun in his right hand. The gun is again some-

Figure 59 ②.

where in the lower back. Your side twist this time is to the *left* instead of to the right, striking your opponent's wrist or forearm with your left elbow (fig. 60 ①). Your left arm is then wrapped behind your opponent's right elbow and pressed on his upper arm with his wrist near your shoulder or neck (fig. 60 ②). You will then grasp your left hand with your right one and press your left forearm against your

FIGURE 59 ③.

opponent's right elbow. A swift twist to the front will bring your opponent to the ground, and if carried to extremes, will break his arm. Note that during the entire operation, the muzzle of the gun is always safely pointed elsewhere (fig. 60 ③).

■ 65. THIRD DEFENSE AGAINST GUN IN BACK (LEFT HAND).—This defense is equally effective when the gun is held in either your opponent's right or left hand. It does not make any

FIGURE 59 ④.

difference whether you know in what hand it is held since your initial move is the same. The description here will be given for the gun held in the left hand, but the action can be duplicated for the right. Figure 61 ① illustrates the original hold-up. The body side twist is executed to the right. Your right elbow strikes your opponent's hand or wrist (fig. 61 ②). You then pivot to the right, and placing your left hand or

FIGURE 60 ①.

wrist against your opponent's left wrist and grasping the barrel of the gun with your right hand, palm upward, you will bring pressure to bear on his hand and trigger finger by pushing the barrel toward his elbow. This will release his hold on the pistol and possibly break the captured index finger (fig. 61 ③). You will now have the gun in your possession and opposite your left shoulder. By reversing strongly

FIGURE 60 ②.

to the right, you can strike the gunman on the chin or neck with the butt of the pistol (fig. 61 ④).

■ 66. FIRST DEFENSE AGAINST GUN IN BACK OF NECK (RIGHT HAND).—This defense is practical only when you are certain that the gun is held in your attacker's right hand. Note that in raising the arms, the elbows are this time shoulder-high (fig. 62 ①). The body pivot is executed to the *left,* your left

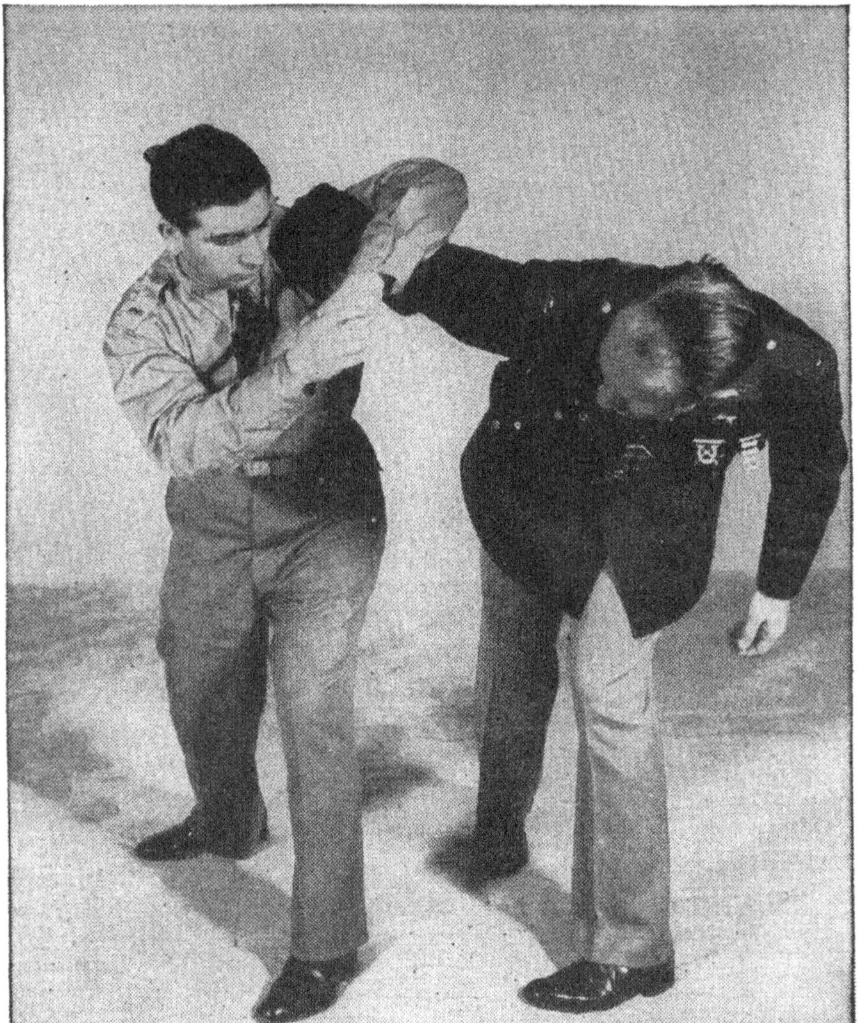

FIGURE 60 ③.

arm coming under your attacker's right elbow (fig. 62 ②). You then reach across with your right hand and grasp your own left hand. Twisting forward and bringing pressure to bear on your opponent's elbow with your left forearm, you can either break the arm or force him to the ground, causing him to release his weapon, which is safely cuddled behind your neck (fig. 62 ⑧).

FIGURE 61 ①.

■ 67. SECOND DEFENSE AGAINST GUN IN BACK OF NECK (RIGHT HAND).—This defense is primarily for an attack with the gun held in the right hand. However, the initial movement can be used for right or left hand. Therefore, the first movement of the defense will be executed even if you do not know in which hand your attacker is holding the pistol. Note that the elbows are held shoulder-high (fig. 63 ①). The body pivot

FIGURE 61 ②.

is executed to the right; at the same time, your right upper arm comes over your opponent's right wrist (if the gun should be in your opponent's left hand, it would come over his left wrist) (fig. 63 ②). You pivot on your right foot and your left foot steps close to your opponent's right foot. Your right upper arm holds your opponent's wrist close to your right

FIGURE 61 ③.

side. Your left arm is crossed *under* your opponent's right upper arm, and your left hand grasps his left lapel. By grasping his right wrist close to your side and lifting with your left upper arm, pressure is brought to bear on your opponent's right elbow, bringing him to his toes and enabling you to march him where you will (fig. 63 ③).

FIGURE 61 ④.

■ 68. THIRD DEFENSE AGAINST GUN IN BACK OF NECK (LEFT HAND).—Note that again the elbows are carried shoulder-high (fig. 64 ①). The initial movement in this defense is identical with that in figure 63 ①. The body twist is executed toward the right and the right arm is brought over your opponent's *left* wrist (fig. 64 ②). You now pivot on your right foot, and

FIGURE 62 ①.

placing your left hand against your opponent's shoulder or upper arm, you bring your right forearm or wrist under your opponent's left elbow and lock it on your own left forearm.

It is now a simple matter to apply the pressure swiftly to cause your attacker to drop his weapon. Severe pressure will result in a broken arm.

FIGURE 62 ②.

■ 69. HELPING A FRIEND IN NEED.—You are standing behind a gunman who has held up a friend, but who is unaware of your presence (fig. 65 ①). Note that as you approach from the rear, your right hand is carried low, and your left one shoulder-high. When you place your hands on your opponent, both hands must grasp simultaneously. The right hand, palm

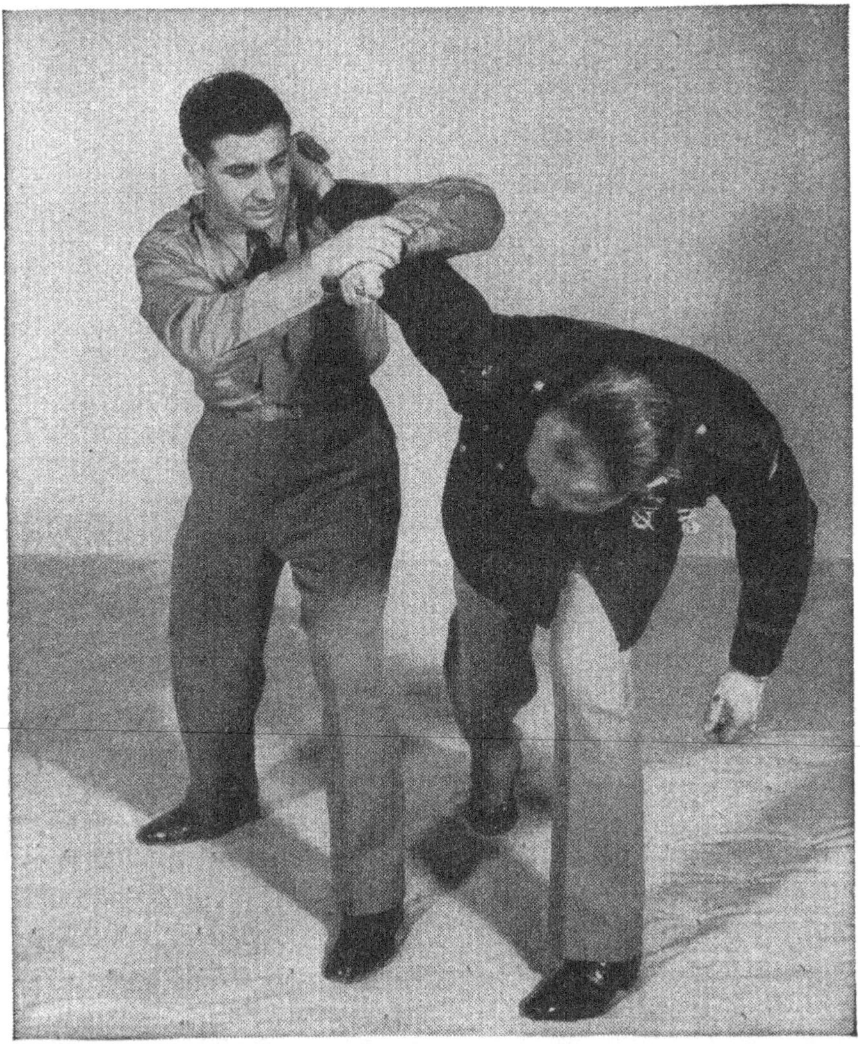

FIGURE 62 ③.

up, lifts on your opponent's hand while your left hand is pressed behind his right upper arm and pushes (fig. 65 ②). By continuing pressure against the upper arm with your left hand and by turning your body to the left and continuing pressure to your opponent's rear against his right hand, you can topple him to the ground or dislocate his shoulder (fig. 65 ③).

FIGURE 63 ①.

■ 70. DEFENSE AGAINST INDIVIDUAL WHO IS HOLDING YOU AND PARTNER AT POINT OF GUN.—Your original position is illustrated in figure 66 ①. The gunman is moving the muzzle of the gun slowly from your friend to yourself and back again. He is attempting to keep you both covered. As the gun swings away from you, step forward with your left foot, and placing your hand on the back of the attacker's gun hand, push strongly to

FIGURE 63 ②.

his left (fig. 66 ②). Now take a step with your right foot, followed by a quick long step with your left one, at the same time twisting the captured hand to your opponent's left, turning it in such a manner that his palm will be turned uppermost and your left armpit will be brought in contact with his right elbow. By bearing down on the right elbow with your armpit and lifting up on the captured hand, excruciating pain is

FIGURE 63 ③.

brought to bear on the gun arm. This will cause him to release his weapon or failing to do so, you will have to break his arm (fig. 66 ③).

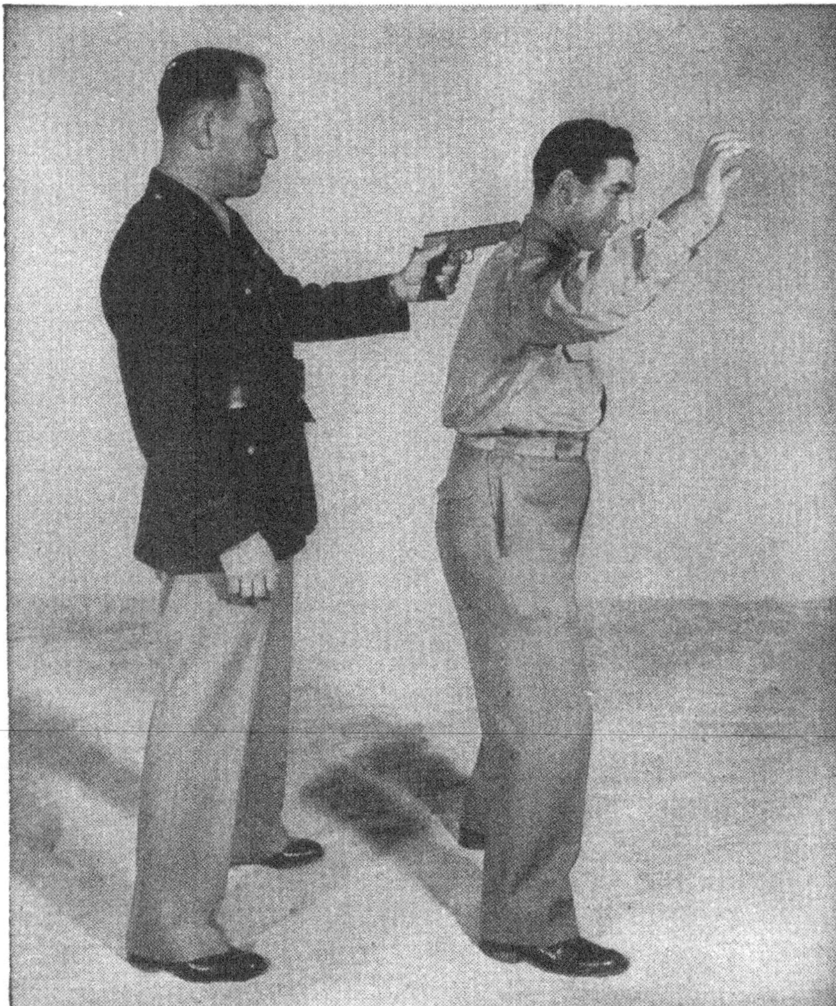

FIGURE 64 ①.

FIGURE 64 ②.

FIGURE 64 ③.

FIGURE 65 ①.

FIGURE 65 ②.

FIGURE 65 ③.

FIGURE 66 ①.

FIGURE 66 ②.

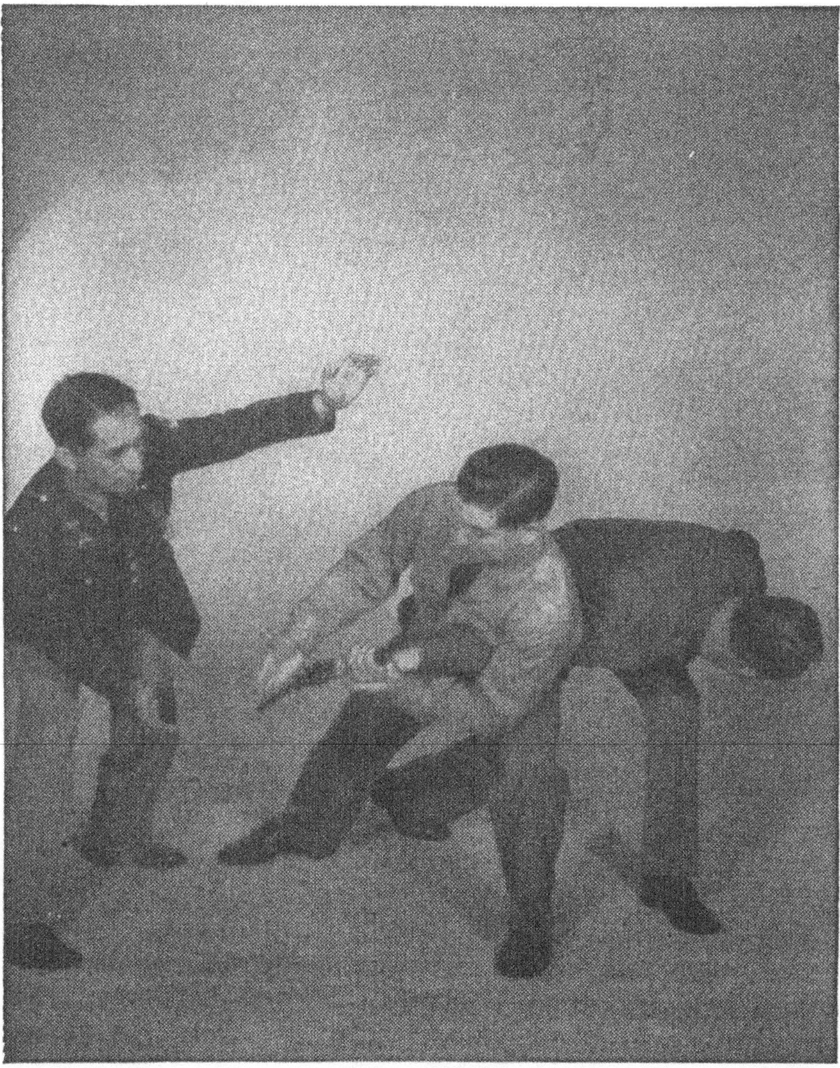

FIGURE 66 ⑧.

Section XI

DEFENSES AGAINST RIFLE

■ 71. **Defense Against Rifle in Front.**—Figure 67 ① illus-

Figure 67 ①.

trates the hold-up. At his order of "Hands up," you start to obey. As your hands reach the level of your shoulders, you immediately execute the body twist to the right as previously described and strike the muzzle of the rifle away from your body with your left forearm or wrist (fig. 67 ②). You imme-

FIGURE 67 ②.

diately follow this movement by stepping forward with your left foot, grasping the rifle over your opponent's left hand with your right one, and grasping his right hand with your left one, thus preventing him from following through with a butt stroke (fig. 67 ③). You are now in excellent position to apply leverage, pulling with your left hand and pushing with your right one, at the same time stepping to your oppo-

FIGURE 67 ⑧.

nent's right with your own right foot. This will take him off balance and at the same time enable you to strike him on the head with the muzzle of the rifle (fig. 67 ④).

■ 72. DEFENSE AGAINST RIFLE IN BACK.—Figure 68 ① illustrates a man holding you up with a rifle in your back. You start to elevate your hands as ordered, and when your hands

FIGURE 67 ④.

reach shoulder height, you execute the side twist to the right, bringing your elbow back so as to strike the muzzle of the rifle. This will deflect it away from your body. Your feet are not moved from their original position. All the twist is made from the hips or waist (fig. 68 ②). You now pivot on your right foot and turn to the right, facing your opponent. Your right arm wraps under the rifle and over your opponent's

FIGURE 68 ①.

left wrist while your left hand is placed on your opponent's right hand where it grasps the stock. This action will prevent his using the butt stroke (fig. 68 ③). You now pull with your left hand and push with your right shoulder and right arm, forcing your opponent to the ground and causing him to release the rifle (fig. 68 ④).

FIGURE 68 ②.

■ **73. DEFENSE AGAINST LONG LUNGE WITH BAYONETED RIFLE.**—Figure 69 ① depicts an unarmed man facing a man armed with a rifle and bayonet. As your attacker executes the long lunge, you execute the body side twist to the right; at the same time strike either the bayonet or the muzzle of the rifle with your wrist or forearm, deflecting it from your body

FIGURE 68 ③.

(fig. 69 ②). Immediately the bayonet passes your body, take a step forward with your left foot and grasp either the bayonet hilt and muzzle or the stock of the rifle over your opponent's left hand, at the same time as you grasp his right hand with your left one where he grasps the stock (fig. 69 ③). Your next action is to pull with your left hand toward

FIGURE 68 ④.

your body, at the same time as you push upward and backward against the rifle with your right hand, simultaneously balancing on your left foot and kicking your attacker behind his right knee joint with the calf of your right leg (fig. 69 ④).

Your attacker loses the rifle and, as he falls, you step backward, turning the rifle so that the blade of the bayonet is in a horizontal position. This will simplify the entrance of the

FIGURE 69 ①.

bayonet between your attacker's ribs when you follow through with a lunge of your own (fig. 69 ⑤).

■ 74. DEFENSE AGAINST SHORT LUNGE WITH BAYONETED RIFLE.—This attack is launched from closer range than the former one. The defense may be the same as for the long lunge with the bayonet. However, this is an additional defense that may be better suited to some individuals. The side body twist is executed to the left, at the same time that

FIGURE 69 ②.

the right forearm or wrist is brought against the barrel of the rifle, deflecting the bayonet from your body (fig. 70 ①). As soon as the bayonet has passed your body, grasp your opponent's left hand where it grasps the stock with your right hand, at the same time as you take a long step to your opponent's right, reaching under the rifle with your left hand

and pressing your left shoulder against the stock. Your right hand will then grasp over his right hand where it grasps the small of the stock (fig. 70 ②). You now pull with your left

FIGURE 69 ③.

hand and push with your right one at the same time as you balance on your left foot and strike your attacker with great force with the calf of your right leg behind the knee joint of his right leg (fig. 70 ③). Your opponent will fall to the ground, losing his grasp on the rifle. Retaining your grasp on the rifle, you will take one step backward and present the bayonet blade on a horizontal plane in position for a short lunge between your opponent's ribs (fig. 70 ④).

FIGURE 69 ④.

FIGURE 69 ⑤.

FIGURE 70 ①.

FIGURE 70 ②.

FIGURE 70 ③.

FIGURE 70 ④.

SECTION XII

DEFENSES AGAINST WRESTLING HOLDS

■ 75. FIRST DEFENSE AGAINST HEADLOCK FROM THE SIDE.—Fig-

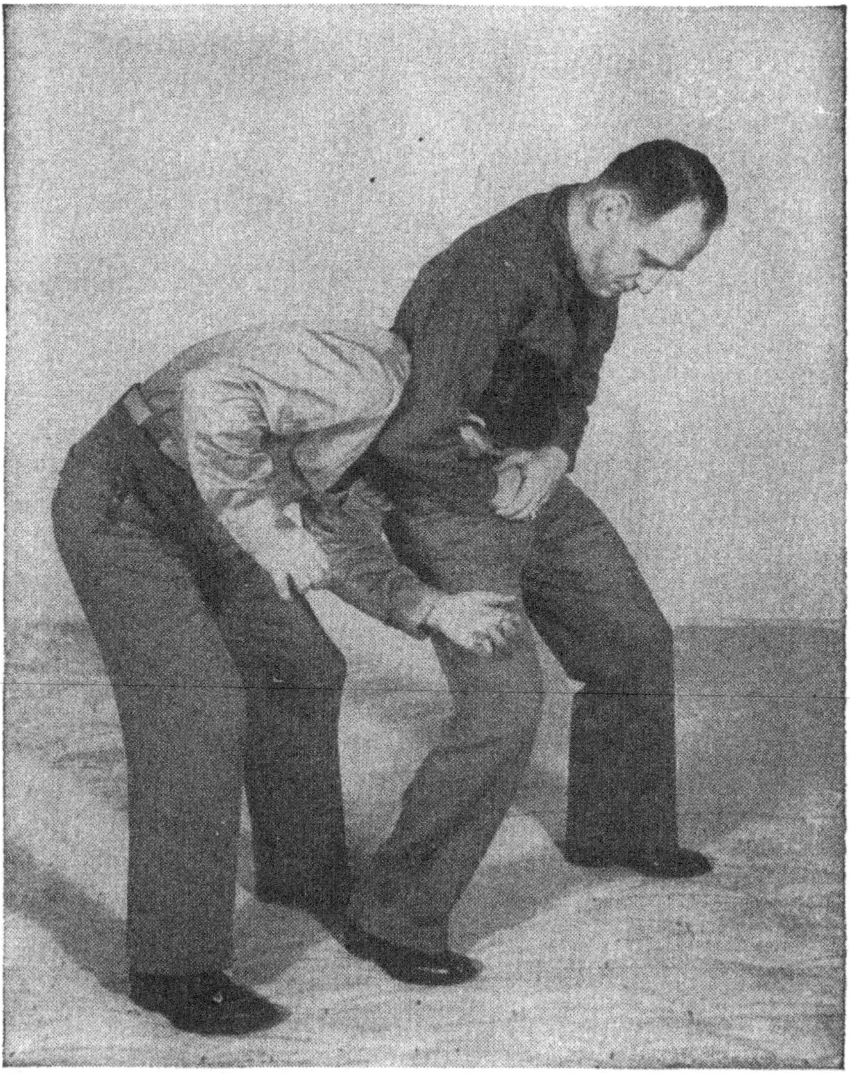

FIGURE 71 ①.

ure 71 ① illustrates the manner in which your opponent will grasp you around the head, locking it close between his arm and chest. You immediately bend your left knee, bringing your knee cap against the back of your opponent's right knee joint, causing his knee to buckle; at the same time you draw

FIGURE 71 ②.

down on his right wrist with your left hand and pull to the right with your right hand grasping his right upper arm, thus making a small space through which to draw your head (fig. 71 ②). Immediately you withdraw your head, bring your opponent's right wrist behind his back with the pressure of your left wrist or hand, then slide your left hand and forearm between the back of your opponent's wrist and grasp it on

FIGURE 71 ③.

the top and rear of his right shoulder. His arm will then be locked in the manner illustrated in figure 71 ③.

463187°—42——15

■ 76. **Second Defense Against Headlock From the Side.**—The original hold is the same as in the defense for the first headlock. However, it is realized that it is not always possible to make a sufficient opening between his arm and body

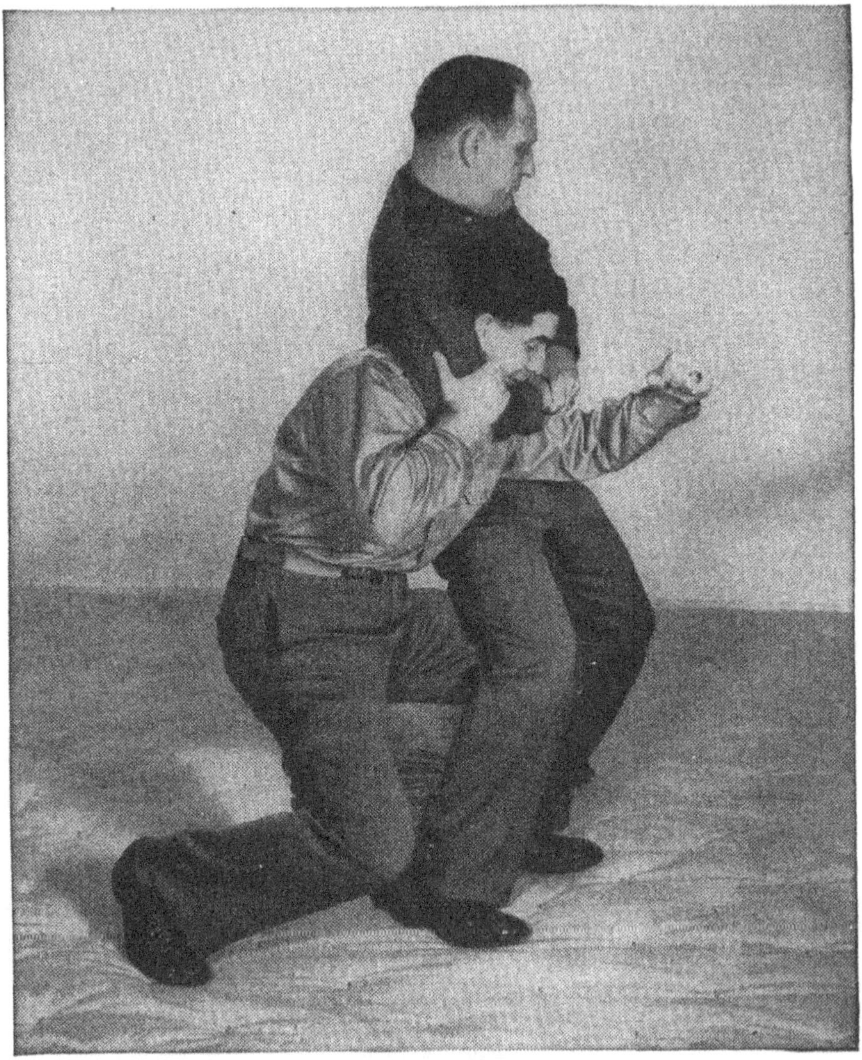

Figure 72 ①.

through which to withdraw your head. Therefore, in the second defense, immediately you cause his knee to buckle in the same manner as in the above defense, bring your left arm and shoulder in front of your opponent's right hip (fig. 72 ①).

Since your attacker is now off balance, your remaining action is to fall over backward, making sure that as you both strike the ground you bring the weight of your left shoulder against

FIGURE 72 ②.

his abdomen. Immediately upon reaching the ground, your left arm is raised and your elbow brought forcibly into your attacker's solar plexus (fig. 72 ②).

■ 77. DEFENSE AGAINST REVERSE HEADLOCK.—When you are attacked in the manner illustrated in figure 73 ①, your immediate reaction is to grasp your opponent's right elbow with your left hand and his right hand with your left hand, your thumb placed on the back of your opponent's right hand over

FIGURE 73 ①.

the center of his knuckles while the fingers of your right hand are wrapped around the little finger edge of his hand. From this position, the hand is turned toward the wrist, causing considerable pain and loosening of the grip. From this position, place your right leg between both your opponent's legs, at the same time that you squat, balancing your weight on your left foot. As you squat, the weight of your body pulls your op-

FIGURE 73 ②.

ponent down and forward, further loosening his grip. This will enable you to bring the captured right hand up and behind your opponent's back as illustrated in figure 73 ③. When you arise from this position, you are usually off balance and will fall backward. You will increase the momentum of your fall by pushing with your right foot and then kicking it upward against your opponent's right knee or into his crotch,

FIGURE 73 ③.

at the same time as you apply pressure on the captured right hand. This will usually result in your attacker's head being crashed into the ground (fig. 73 ③). If he attempts a rolling fall and thus avoids making contact with his head against the ground, you will still retain your original grip on arm and hand with which you can control him.

■ 78. ESCAPE FROM HAMMERLOCK.—Figure 74 ① illustrates a hammerlock of a type from which it is very difficult to escape. Your immediate reaction on realizing that an attacker is at-

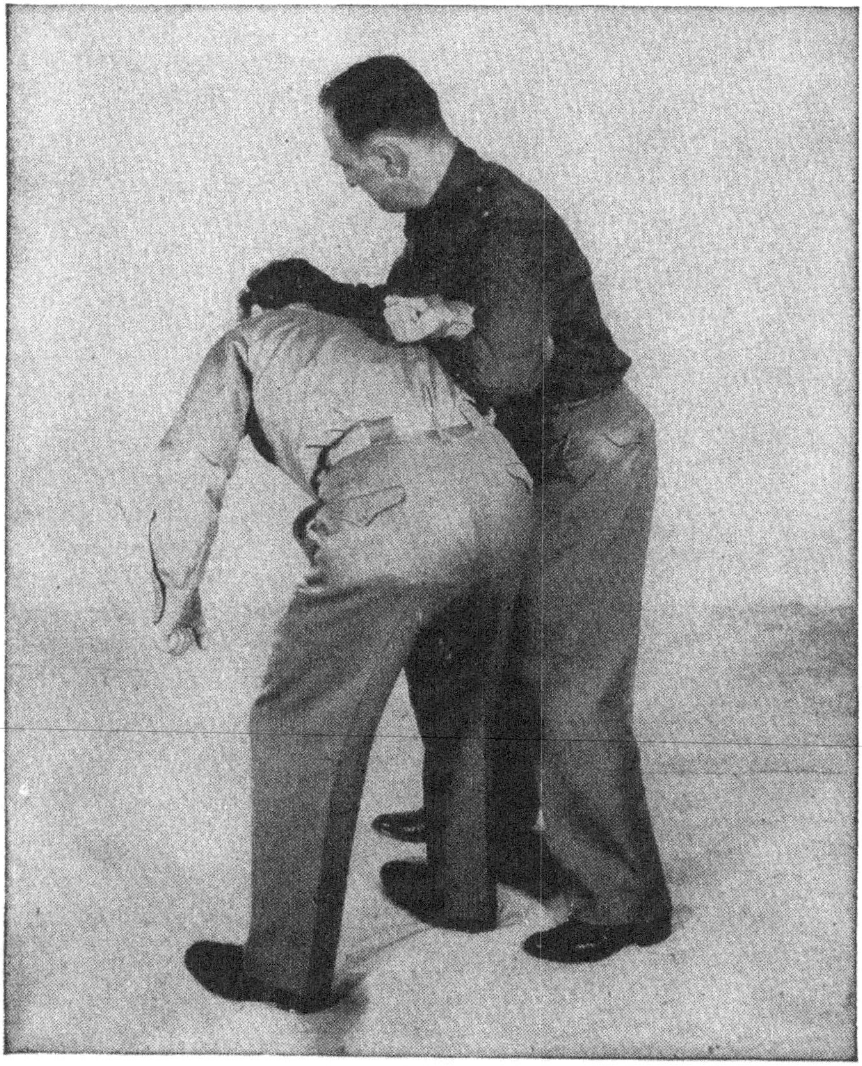

FIGURE 74 ①.

tempting to place a hammerlock on you is to execute three simultaneous actions. The knees are bent, your body is bent from the waist, and the captured arm is extended in as straight a line as possible (fig. 74 ②). This will free your arm from its imprisonment. You immediately straighten your legs and body and throw your extended right arm, which is probably placed on your opponent's left shoulder, and pass this arm

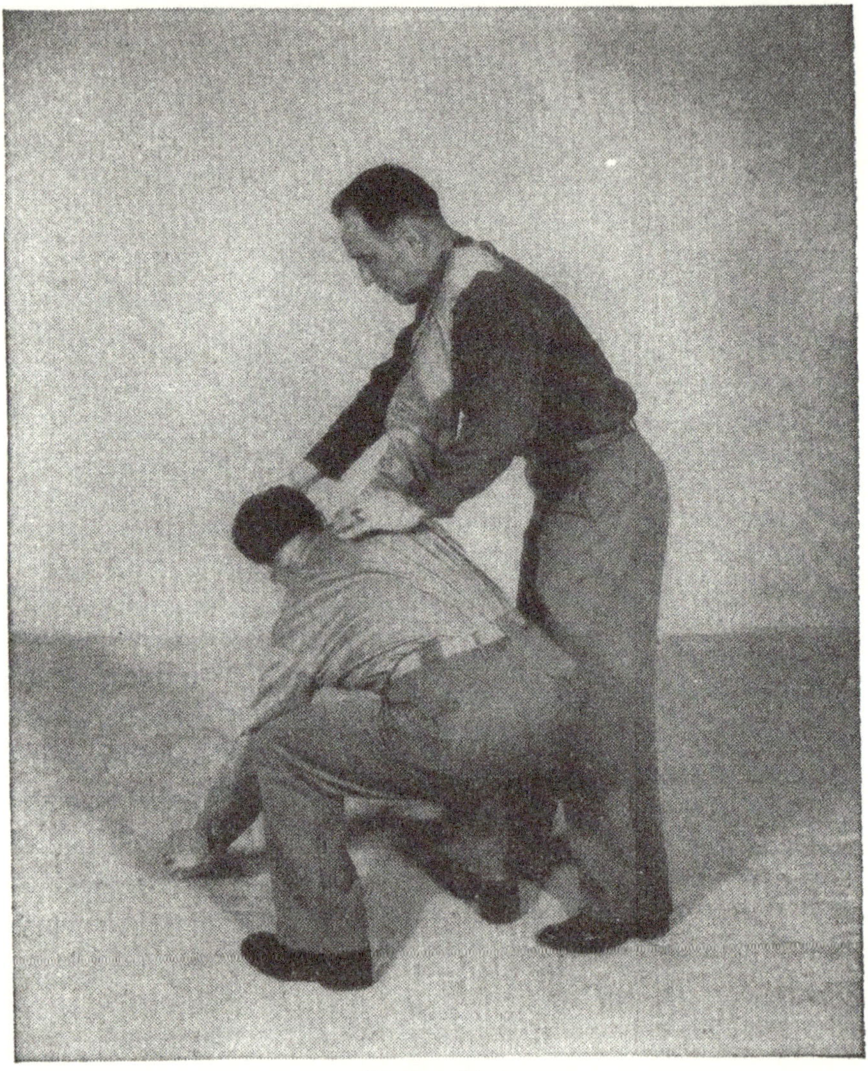

FIGURE 74 ②.

over the back of your attacker's neck, at the same time as you place your right foot to the right of your opponent's right foot and pointing in the same direction (fig. 74 ③). It is now a simple matter to overbalance your opponent over your hips by simply pivoting to the left and pulling strongly with your right arm across the back of his neck (fig. 74 ④).

■ 79. DEFENSE AGAINST FLYING TACKLE.—It is usually a diffi-

FIGURE 74 ③.

cult thing to avoid a flying tackle executed by a skilled athlete. Your best defense is to permit him to get as close as possible and then raise your knee quickly so as to bring your knee cap in violent contact with some part of his face (fig. 75).

■ 80. DEFENSE AGAINST FULL NELSON.—Figure 76 ① illustrates the hold known to the wrestling profession as the full nelson. It is usually a dangerous hold when executed by a well-trained

FIGURE 74 ④.

wrestler. Your immediate reaction is to leap into the air, permitting your weight to be supported by your attacker's arms, twist to the left, and bring your left leg behind your opponent's right leg (fig. 76 ②). Immediately upon striking the ground behind your opponent, drive your right knee forcibly against the back of your opponent's right knee joint (fig. 76

⑧), and immediately fall to add momentum to your left arm. As you strike the ground, your left elbow makes contact with your opponent's throat or chin.

FIGURE 75.

■ 81. DEFENSE AGAINST FLYING MARE.—Figure 77 ① illustrates another hold well-known to the wrestling gentry. Apparently in this illustration, the man being attacked is lost. Immediately you realize that your opponent is attempting a flying mare, you bring the thumb of your free hand, which in this case happens to be the left one, and dig deeply into the nerve tissues just above your opponent's hip bone (fig. 77 ②). While this spot is not necessarily sensitive in all individuals, it will cause loss of motor power in the legs of any person.

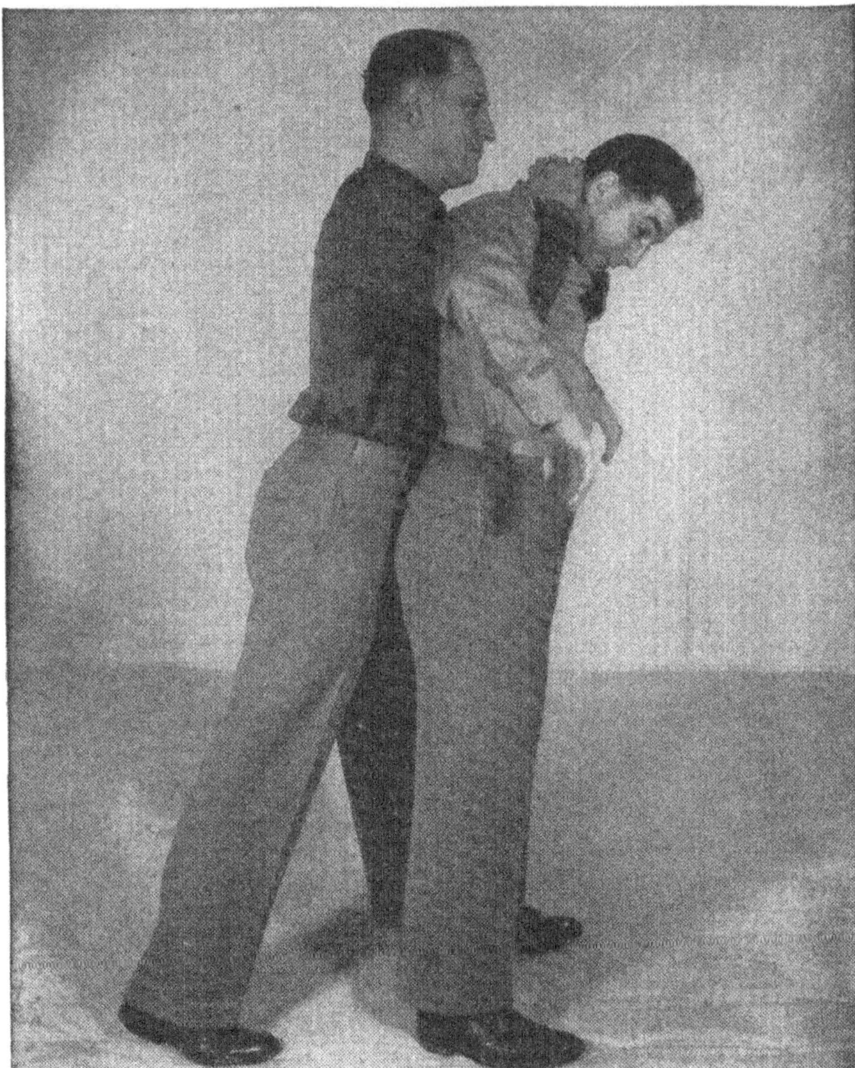

FIGURE 76 ①.

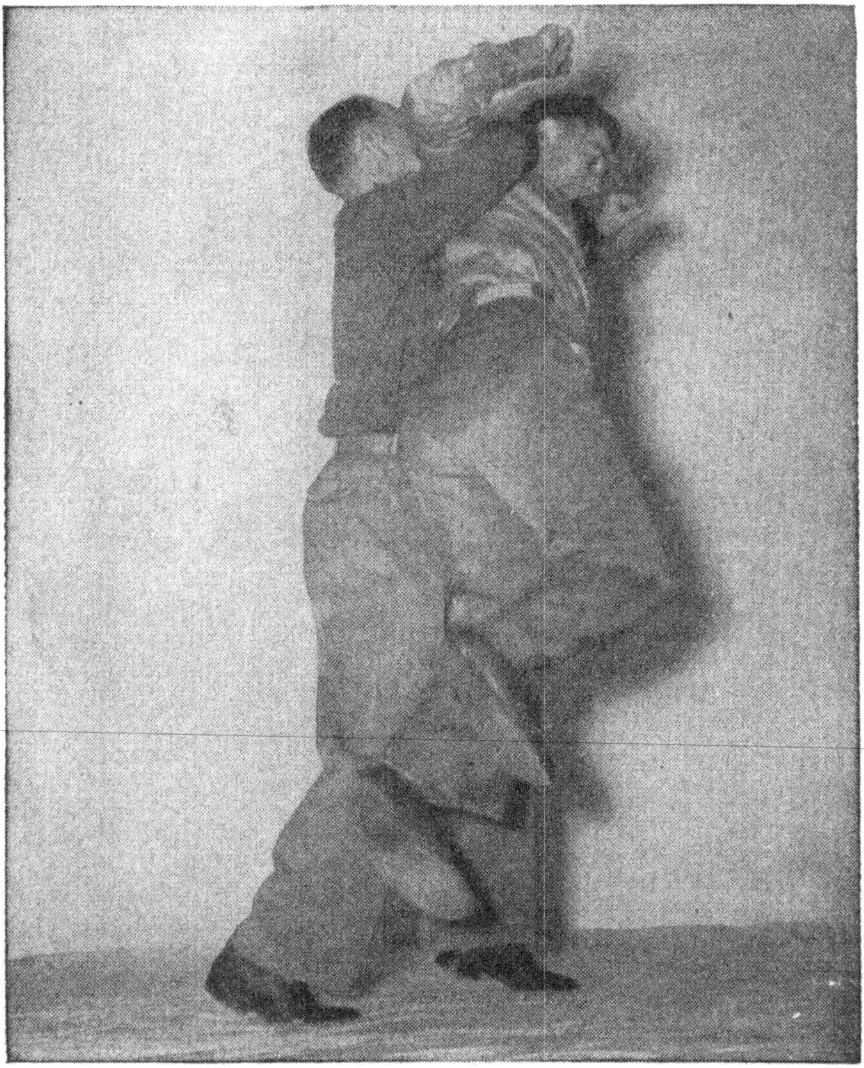

FIGURE 76 ②.

FIGURE 76 ⑧.

FIGURE 77 ①.

FIGURE 77 ②.

SECTION XIII

DEFENSES AGAINST GRIPS ON GARMENTS OR HAIR

■ 82. DEFENSE AGAINST COLLAR HOLD FROM REAR.—Figure 78 ①

FIGURE 78 ①.

illustrates the situation where your opponent has grasped your collar from the rear with his right hand and is starting to pull you backwards. As you feel the pull on the collar, do not attempt to resist it, but rather use it to help you in pivoting on your left foot and turning to the right. As you turn, place your left hand on the wrist of the attacking arm of your op-

FIGURE 78 ②.

ponent in order to prevent him from loosening his hold on your collar (fig. 78 ②). In the same motion, strike the back of your opponent's right knee joint with the calf of your right leg simultaneously with a straight-arm blow with the heel of your right hand against your opponent's chin (fig. 78 ③).

FIGURE 78 ③.

■ 83. DEFENSE AGAINST HOLD ON BELT.—Figure 79 ① illustrates the original hold. Since the hold in this case is on the left side of the belt, the right hand is brought upward, palm uppermost, striking your attacker's right wrist, releasing the hold from the belt. Immediately his hand is removed from the belt, follow through by gripping tightly with your right hand on the wrist and placing the fingers of your left hand

FIGURE 79 ①.

against the knuckles of your opponent's right hand and applying pressure in the direction of his wrist (fig. 79 ②). Continuous pressure in this manner will bring your opponent's hand up his back on the right side, and you can further bring him to his toes in a very painful position by pressure under his chin with your right hand (fig. 79 ③).

FIGURE 79 ②.

■ 84. DEFENSE AGAINST CHEST PUSH.—Figure 80 ① illustrates the manner in which a great many fights are started. One individual pushes the other on the chest. The individual pushed feels that he has to retaliate, and so it goes until somebody starts throwing punches. Immediately your opponent's hand rests on your chest, place both hands over his hand, which in this case happens to be the right one (fig. 80 ②).

FIGURE 79 ③.

Holding his hand tightly against your chest, bend swiftly from the waist and slightly at the knees (fig. 80 ③). This will usually bring your attacker to the ground with very little effort on your part. However, if you feel it necessary to take him a prisoner, you will not take him all the way to the ground, but will simply make him buckle at the knees; keeping the pressure of your right hand on the back of his right

FIGURE 80 ①.

one, you will place your left hand on the back of his right elbow (fig. 80 ④), and by applying pressure, prevent him from removing his hand. You will now bend your knees to as great a degree as possible and take a long step with your right foot to your opponent's right rear (fig. 80 ⑤). As soon as your right foot strikes the ground, you will balance on it and turn to the left, which will bring you behind your oppo-

FIGURE 80 ②.

nent and facing in the same direction as he is. His hand is still retained against your chest and the hold on his elbow is also retained. His arm will be at right angles to his back.

It is now a simple matter to shift your right hand to his upper arm and your left one to his hand and control him by pressure up his back (fig. 80 ⑥).

FIGURE 80 ⑧.

■ 85. DEFENSE AGAINST GRIP ON FRONT OF GARMENT, THUMB OUT.—When your shirt or tie is gripped with the thumb out as illustrated in figure 81 ①, you will immediately reach up with your right hand, placing the four fingers around the base of your attacker's thumb along the back of his hand, while the entire first and second joints of your thumb are placed under the inside of his thumb and pressed backward

FIGURE 80 ④.

(fig. 81 ②). With this hold your opponent can be brought to the ground with very little effort.

■ 86. DEFENSE AGAINST GRIP ON FRONT OF GARMENT, THUMB IN.—If your shirt or tie is gripped with the thumb turned inward as illustrated in figure 82 ①, you grip the outside of the base of your opponent's thumb with your four fingers and place your first and second fingers over your opponent's fingernail and attempt to force the first joint of his thumb

FIGURE 80 ⑤.

toward the third joint. This will cause excruciating pain and very sudden release of any grip your opponent may have had (fig. 82 ②).

■ 87. DEFENSE AGAINST HAIR HOLD FROM FRONT WITH RIGHT HAND.—Reach upward with your right hand, placing your fingers around the little finger edge of his hand and grasping the palm while the thumb is placed on the back of the

FIGURE 80 ⑥.

knuckles (fig. 83 ①). Holding his hand close to your head, you now execute a half right-face and bending from the waist get your head as close to the ground as possible (fig. 83 ②). This will bring considerable strain on your attacker's wrist and possibly cause him to release his hold. As soon as his hold loosens, you will reinforce your original right-hand hold with a similar grip with your left hand which will mean that

FIGURE 81 ①.

both your thumbs will be crossed over the back of his knuckles and the fingers of both hands will meet in the palm of the hand. By applying pressure on the hand toward his wrist, he will be forced to turn and walk where you will (fig. 83 ③).

■ 88. **Defense Against Hair Hold From Rear.**—Figure 84 ① illustrates the original attack. Before turning your body, your right hand reaches upward and is placed over your op-

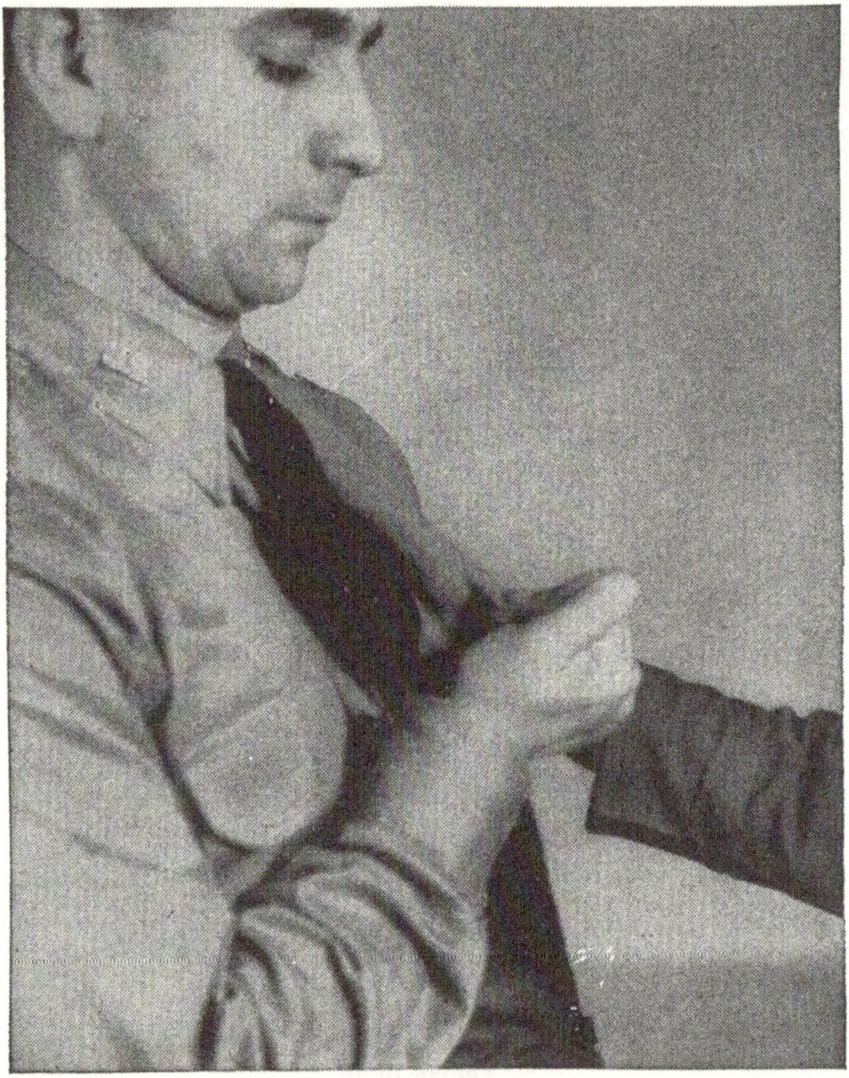

Figure 81 ②.

ponent's right hand, your four fingers wrapped around the base of his thumb and your thumb on the back of his hand. You now execute a half right-face, bending from the waist, getting your head as close to the ground as possible. The strain on your opponent's wrist will cause him to loosen his grip (fig. 84 ②). You will now pivot on your right foot, fac-

FIGURE 82 ①.

ing your opponent, and reinforcing your original hold with a similar hold with your left hand, twist powerfully to your left. Your opponent will either execute a complete somersault or suffer a broken wrist (fig. 84 ③).

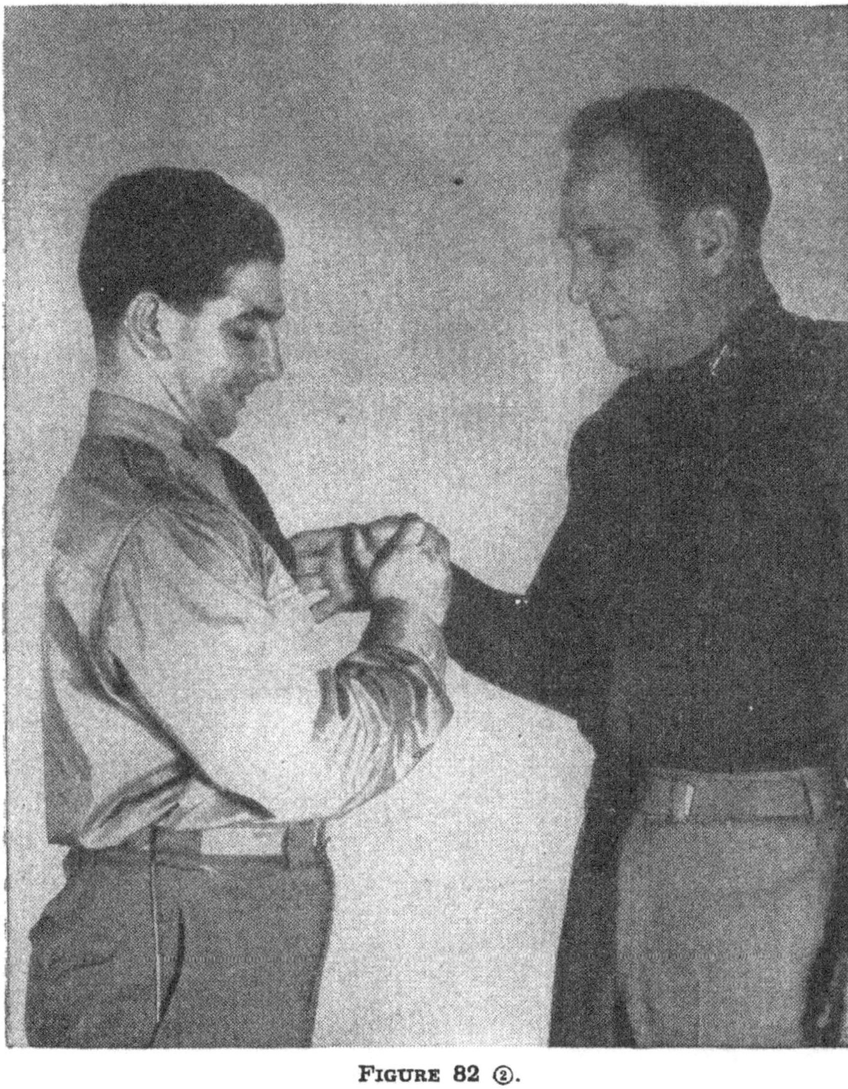

FIGURE 82 ②.

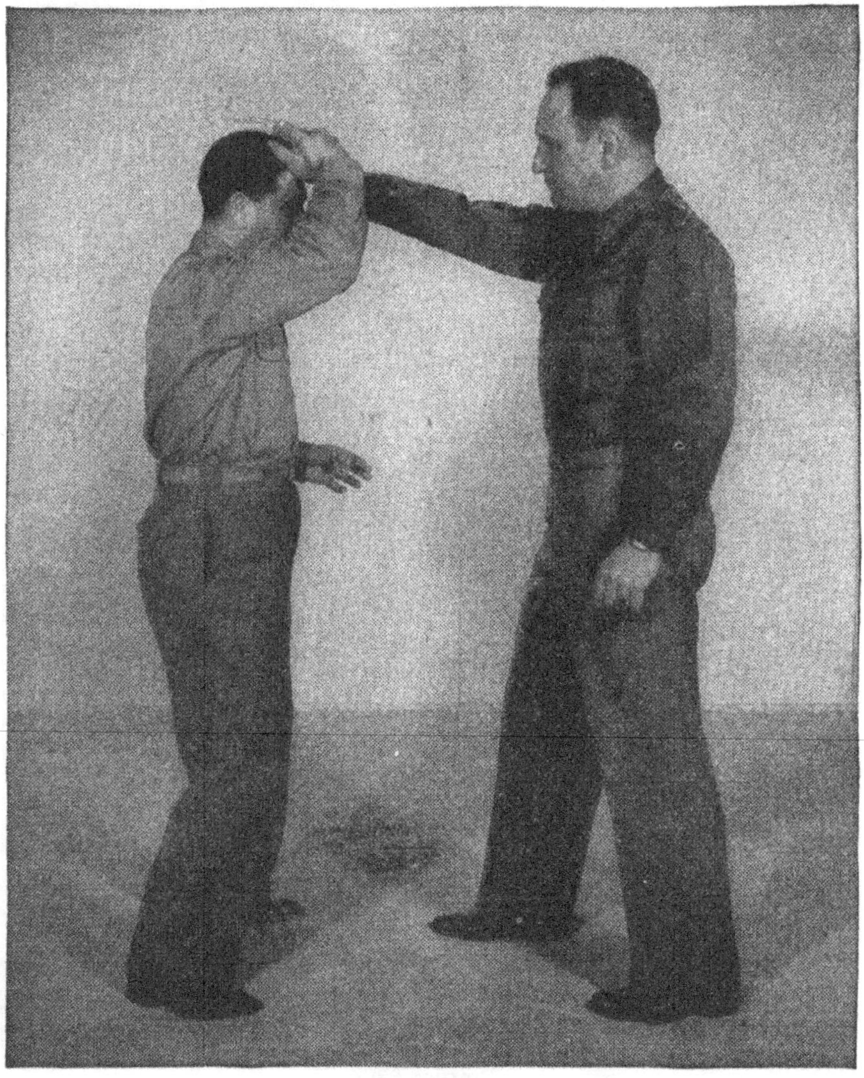

FIGURE 83 ①.

FIGURE 83 ②.

FIGURE 88 ⑧.

FIGURE 84 ①.

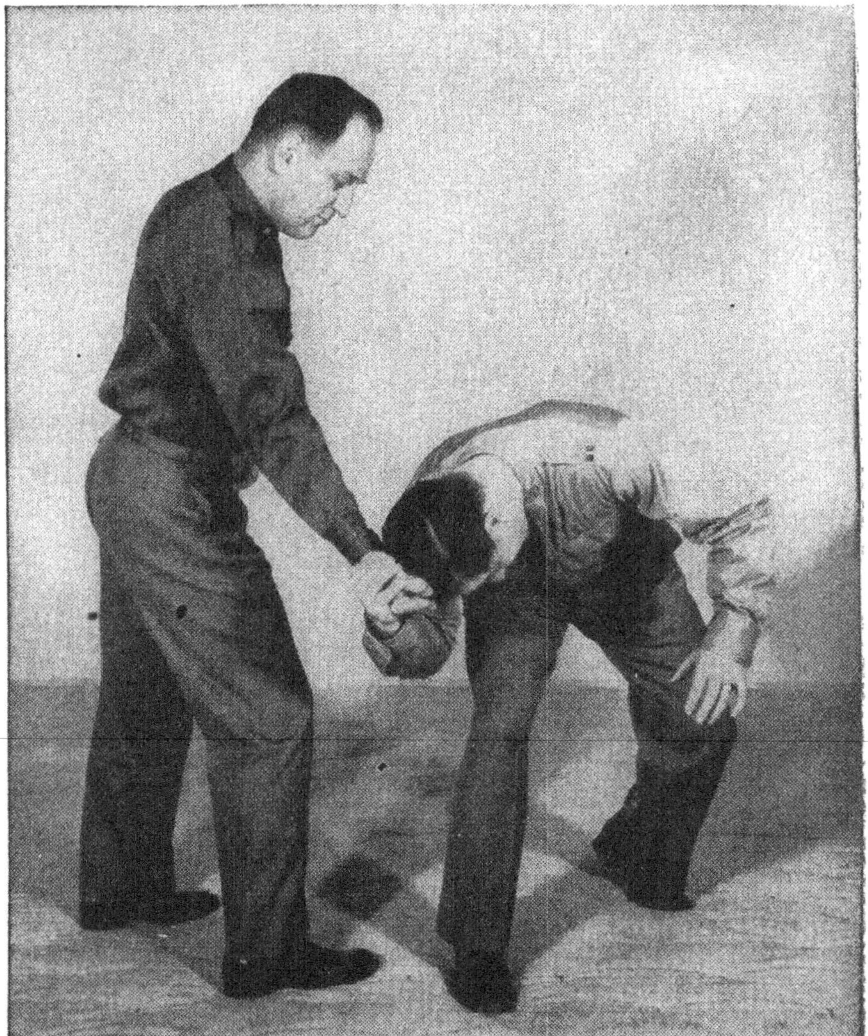

FIGURE 84 ②.

FIGURE 84 ③.

SECTION XIV

DEFENSES AGAINST FIST ATTACKS

■ 89. APPROACHING CLEVER BOXER.—When your attacker as-

FIGURE 85 ①.

sumes boxing position, illustrated in figure 85 ①, you can be fairly certain that he is skillful in the use of his fists, and you square off in a similar manner, keeping your hands open.

When you get just out of reach of a possible left jab, you will suddenly launch your body through the air at your opponent, turning toward the left as your spring. Your lower

FIGURE 85 ②.

left leg will strike behind your opponent's left knee joint while the back of your right leg will catch him across the center of his body (fig. 85 ②). Your left hand will be used to break the force of your fall and your right one to ward off a possible jab by your attacker. This scissoring action on your part will bring your opponent to the ground with great force. You can immediately follow through by lifting your

FIGURE 85 ③.

right foot and bringing the heel down into your opponent's midsection.

■ 90. DEFENSE AGAINST LEFT JAB TO HEAD.—When your opponent throws a left jab to your head, you break the blow with the outside of your right forearm, keeping your left hand and arm extended to ward off a possible right cross (fig. 86 ①). You then take a short step forward with your right foot and shifting your balance to it, raise your left leg

FIGURE 86 ①.

and bring the calf against the back of your attacker's left knee joint, at the same time as you bring the heel of your left hand against his chin (fig. 86 ②).

■ 91. **Defense Against Left Hook to Head.**—When the left hook is brought to the head, the head is drawn backward just far enough to avoid the blow, and the *right* forearm is

Figure 86 ②.

lifted so as to contact your opponent's left wrist or forearm while your right hand is placed on the back of his elbow (fig. 87 ①). By continuing the momentum of his blow toward your left, his left hand and wrist will slide naturally into the crook of your left elbow while your right hand pushes strongly against his left elbow. It is now a simple matter to lock both

FIGURE 87 ①.

your hands over that joint (fig. 87 ②). This can be used as a come-along as well as an arm breaker.

■ 92. DEFENSE AGAINST LEFT HOOK TO BODY.—The blow is stopped with the little finger edge of your right forearm, which is presented strongly against your opponent's left forearm. Your left hand is raised as a defense against a possible follow through with your opponent's right hand (fig. 88 ①).

FIGURE 87 ②.

You now reach across with your left hand, grasping the back of your opponent's left elbow and pulling the elbow toward you, at the same time as you press to his rear with your right forearm against his left wrist (fig. 88 ②). Your right arm is then brought under your attacker's left armpit and over his shoulder; simultaneously with this action, you pivot on your left foot, placing your right foot to the right of your

FIGURE 88 ①.

attacker's right foot and pointing in the same direction as his. You immediately reinforce the right hand on his shoulder with your left hand (fig. 88 ③). By striking backward with your hips and twisting your body forward and to the left, your opponent will be propelled over your hips.

FIGURE 88 ②.

■ 93. DEFENSE AGAINST RIGHT CROSS TO HEAD.—When a right cross is thrown to your head, your arms are crossed at the forearms with your left arm being closest to your face, catching your opponent's blow in the crook or V formed by your forearms (fig. 89 ①). As soon as the blow is stopped, your left hand is turned with the thumb toward your opponent's body and the fingers grasping his forearm and you take a long step

FIGURE 88 ③.

with your right foot, placing it to the right front of your opponent's right foot, and immediately grasp his upper arm from the outside with your right hand (fig. 89 ②). You immediately pivot to the left, bringing your right shoulder under your attacker's right armpit, retaining both holds on his arm. By bending strongly from the waist, your opponent is catapulted over your head (fig. 89 ③).

FIGURE 89 ①.

■ 94. DEFENSE AGAINST RIGHT HOOK TO HEAD.—When the right hook or roundhouse to the head is thrown, the blow is blocked with your left forearm, and your right arm is raised to ward off a possible blow from the left hand (fig. 90 ①). You now pivot on your left leg to the left, placing your right foot on the outside of your attacker's right foot and pointing in the same direction as his. Simultaneously with this movement,

FIGURE 89 ②.

your right arm is slipped over your attacker's left shoulder and over the back of his neck (fig. 90 ②). Your left hand then reaches across and grasps the outside of your attacker's right upper arm. You now bend swiftly from the waist, at the same time pulling strongly with your left hand and with your right arm (fig. 90 ③).

FIGURE 89 ③.

■ 95. DEFENSE AGAINST RIGHT HOOK TO BODY.—The blow is stopped by presenting your left forearm forcibly against the thumb side of your attacker's right wrist or forearm (fig. 91 ①). Immediately on stopping the blow, your right hand grasps your attacker's right wrist and pulls him toward you (fig. 91 ②). You now step to your opponent's right rear with your left foot, at the same time wrapping your left arm around

FIGURE 90 ①

your attacker's right arm and locking your hand on your chest (fig. 91 ③). In this position, extreme pressure can be brought to bear on his elbow joint.

■ 96. Defense Against Uppercut.—Your defense against the uppercut is to bring the little finger edge of the hand nearest the attacking arm forcibly against the base of your attacker's

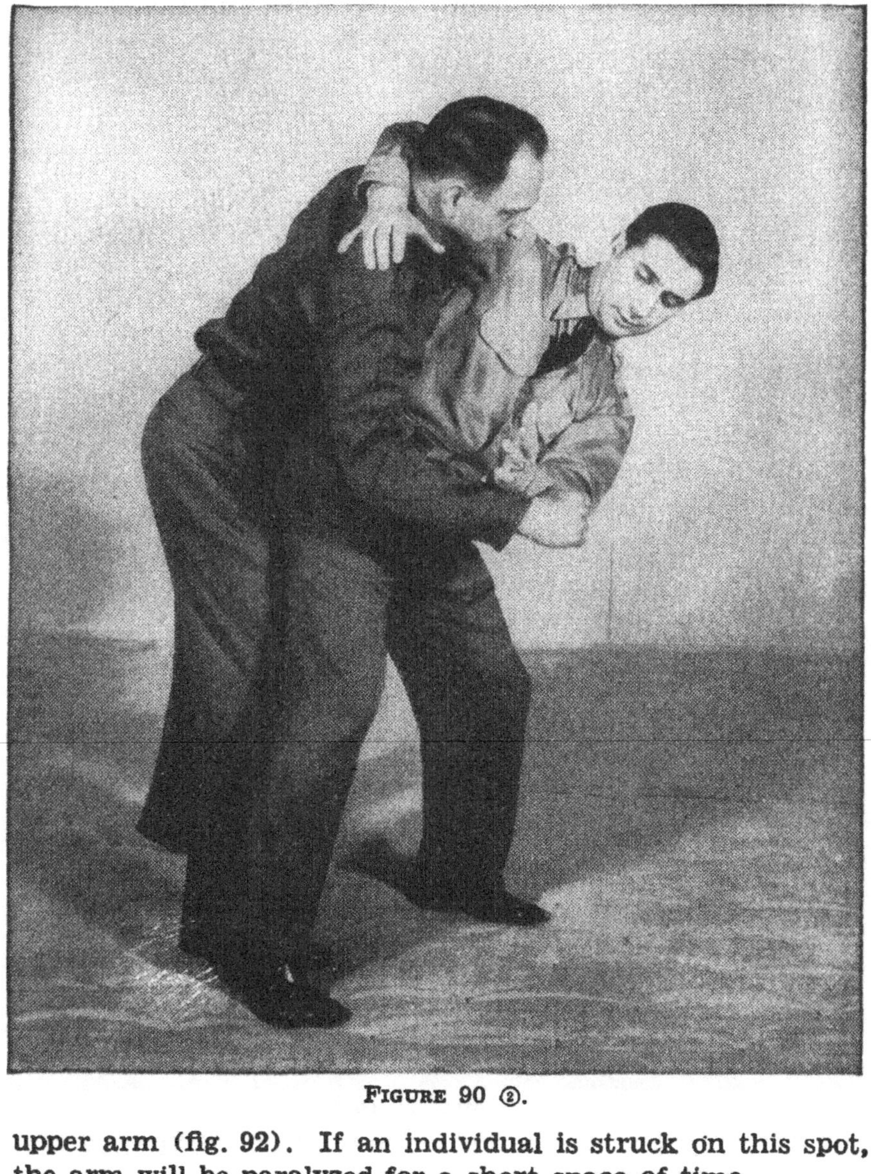

Figure 90 ②.

upper arm (fig. 92). If an individual is struck on this spot, the arm will be paralyzed for a short space of time.

■ 97. How To Stand When Expecting Trouble.—Figure 93 illustrates the proper stance for an individual who is expecting trouble from someone with whom he is conversing. Note the manner in which the hands are clasped. In this position,

Figure 90 ③.

you have a perfect all-around defense for an unexpected blow by your opponent using either hand.

■ 98. Defense Against Sudden Right-Hand Blow.—As your opponent launches his sudden right-hand blow to your chin your hand is cupped near the chin so as to catch his fist, the same time as you lean slightly backward, giving with the blow. Immediately upon stopping the blow, your left hand

FIGURE 91 ①.

from its position on your right upper arm is cut quickly against your opponent's neck, striking with the little finger edge (fig. 94 ②), and continued on its course, moving your opponent's right hand to your left. This opens the way for a cut with the little finger edge of your right hand against the right side of your opponent's neck (fig. 94 ③). Figure 94 ④ illustrates the actual contact of the blow and shows

FIGURE 91 ②.

your left hand raised for a possible finishing touch. These blows are struck in rapid succession; any one of the three is a potential knockout.

■ 99. **Defense Against Sudden Left-Hand Blow.**—As your opponent throws a sudden left-hand blow, your cupped left hand catches it close to your chin as you bend your body

Figure 91 ③.

slightly backward in order to give with the blow (fig. 95 ①). Immediately upon stopping the blow, your right hand, which was clasped on your left upper arm, is cut quickly to your right, striking your opponent in the solar plexus with the tips of your fingers (fig. 95 ②). This blow is immediately followed by a strong cross-cut to the left side of your opponent's neck with the little finger edge of the left hand

FIGURE 92.

(fig. 95 ③). Your right hand, which was waist-high, is then brought diagonally upward, striking your opponent's throat with the little finger edge of the hand while your left hand is cocked for a possible finishing touch (fig. 95 ④).

■ 100. **First Method of Holding Individual Whom You Are Trying To Quiet, But Who Might Suddenly Strike.**—When you suspect that an individual whom you are trying to quiet

Figure 93.

might strike you without warning, you hold him loosely by his left elbow with your right hand (fig. 96 ①). You are on the alert for a possible blow and immediately that he starts a blow with either hand, you swing him to your left, using your right hand gripping his left elbow as a turning lever. This will expose his left side for a cross-cut by the little finger edge of your left hand (fig. 96 ②). The blow makes contact

FIGURE 94 ①.

in the soft spot between your attacker's hip bone and his short ribs.

■ 101. SECOND METHOD OF HOLDING INDIVIDUAL WHOM YOU ARE

TRYING TO QUIET, BUT WKO MIGHT SUDDENLY STRIKE.—The method illustrated in figure 97 ①, grasping your opponent's left wrist with your right hand, is not as safe as the first

FIGURE 94 ②.

method. However, if he should suddenly strike with his right hand, you move his left arm to your left so as to catch his right-hand blow (fig. 97 ②). Immediately his blow is stopped, you grasp the projecting right wrist with your left hand with the palm of your hand turned to the left and the thumb down. (Study illustration.) You now pivot to the left on your left foot, placing your right foot on the outside

FIGURE 94 ③.

of your opponent's right foot, and face in the same direction as he is facing. You bring both his arms over your right shoulder (fig. 97 ③). From this position, it is a simple matter to throw your opponent over your head simply by drawing on his wrist and bending at the waist.

FIGURE 94 ④.

■ 102. BEATING OPPONENT TO THE BLOW.—When you are caught with your hands at your sides and close to an opponent who suddenly raises his fist to strike you, your quickest defense is the blow illustrated in figure 98. Your right hand is moved in a straight line from its position at your side diagonally upward, striking your opponent between the chin and adam's

FIGURE 95 ①.

apple with your thumb on one side of the throat and your fingers on the other. It is a knock-out and it is done faster in this fashion than in any other.

FIGURE 95 ②.

FIGURE 95 ③.

FIGURE 95 ④.

FIGURE 96 ①.

FIGURE 96 ②.

FIGURE 97 ①.

FIGURE 97 ③.

FIGURE 97 ⑧.

FIGURE 98.

SECTION XV

INCAPACITATING AN OPPONENT

■ 103. TAKING MAN DOWN FROM REAR.—Figure 99 ① illus-

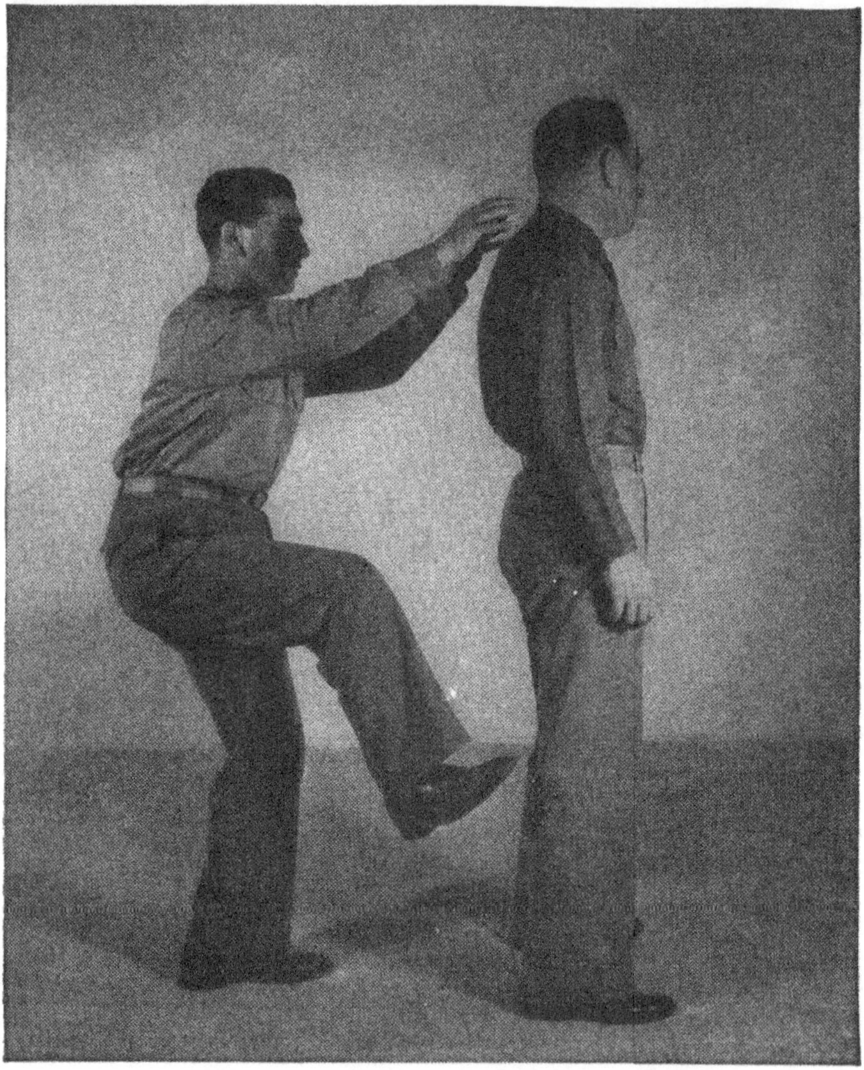

FIGURE 99 ①.

trates the start of the proper method of bringing a dangerous opponent to the ground from the rear. Note that the hands and foot are making contact simultaneously. The hands strike down on the shoulders and pull backward at the same instant as the sole of the right foot kicks strongly against the back of your victim's knee joint (fig. 99 ②). This

FIGURE 99 ②.

attack will drop your victim flat on the ground in an instant.

■ 104. KIDNEY STROKE AGAINST LOW ATTACK.—When an opponent attempts to grasp you around the legs, you stop his forward action with your left hand against his right shoulder, at the same time as you cut across his right kidney with the little finger edge of your right hand (fig. 100).

FIGURE 100.

■ 105. OCCIPITAL STROKE AGAINST LOW ATTACK.—Your opponent in this case is attacking in the same manner. However, if he is dangerous, you will use the occipital stroke at the base of the neck with the little finger edge of your right hand. The stop is made with your left hand as before described (fig. 101).

FIGURE 101.

■ 106. CLAVICLE STROKE (BREAKING COLLAR BONE.)—If the collar bone is struck at the back of its arch with the little finger edge of either hand in the manner illustrated in figure 102, it will result in a clean break.

■ 107. Simple Front Choke.—This again is a defense for a low attack. The momentum of your opponent is again slowed with your left hand against his right shoulder while your right

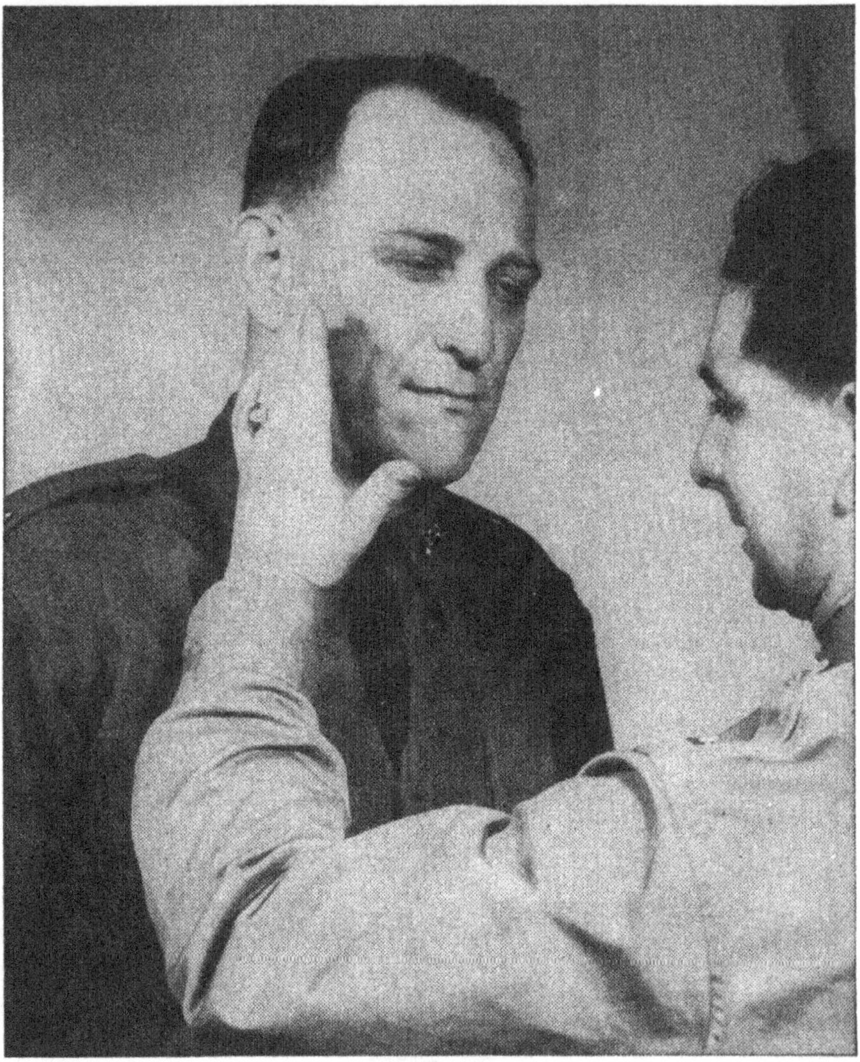

Figure 102.

forearm is slipped under your attacker's throat. His head is midway between your armpit and your chest (fig. 103 ①). As your hand reaches your attacker's throat, it is clasped on your left wrist. Pressure is applied by leaning backward and lifting with your right forearm (fig. 103 ②). A more advanced choke and one which acts more swiftly is the one illustrated in figure 103 ③. Instead of locking the hand, the

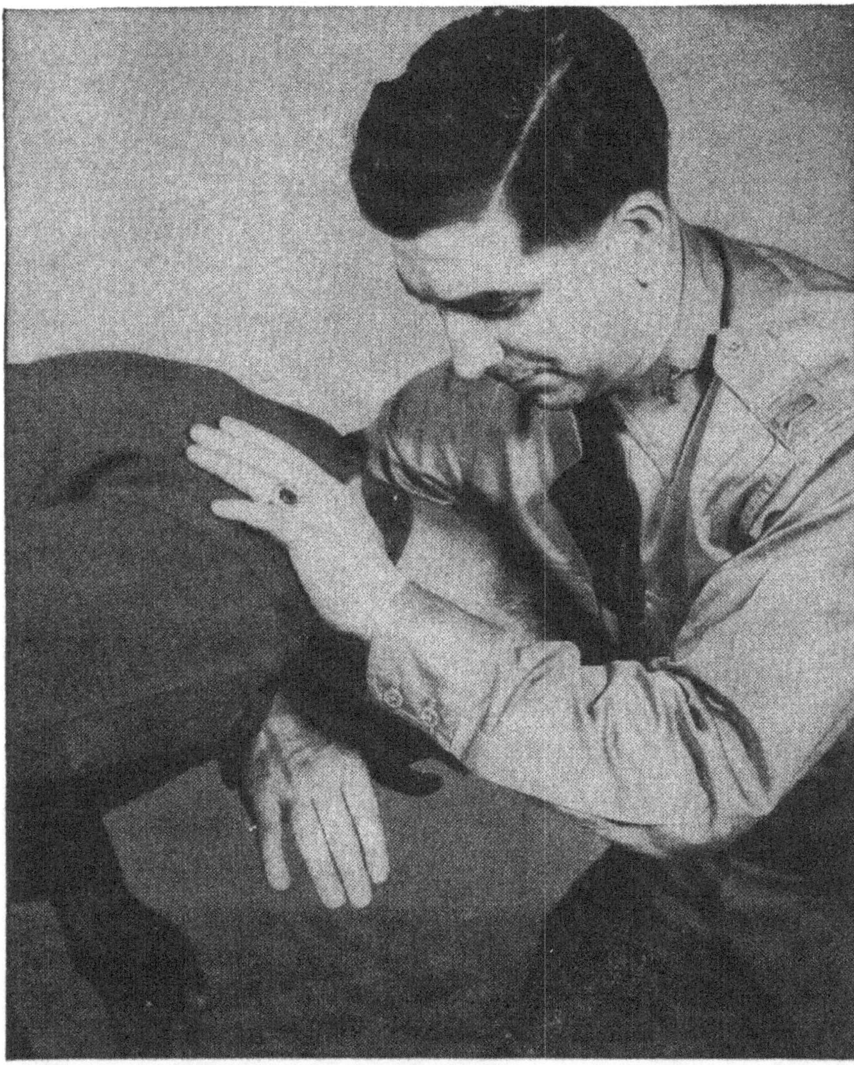

FIGURE 103 ①.

fingers of your left hand grasp the little finger edge of your right hand and pull strongly toward your chest, at the same time as you lean backward.

■ 108. SIMPLE REAR CHOKE.—The rear choke is utilized from the position of the rear take-down previously illustrated in figure 99 ① and ②. As your victim falls backward, your

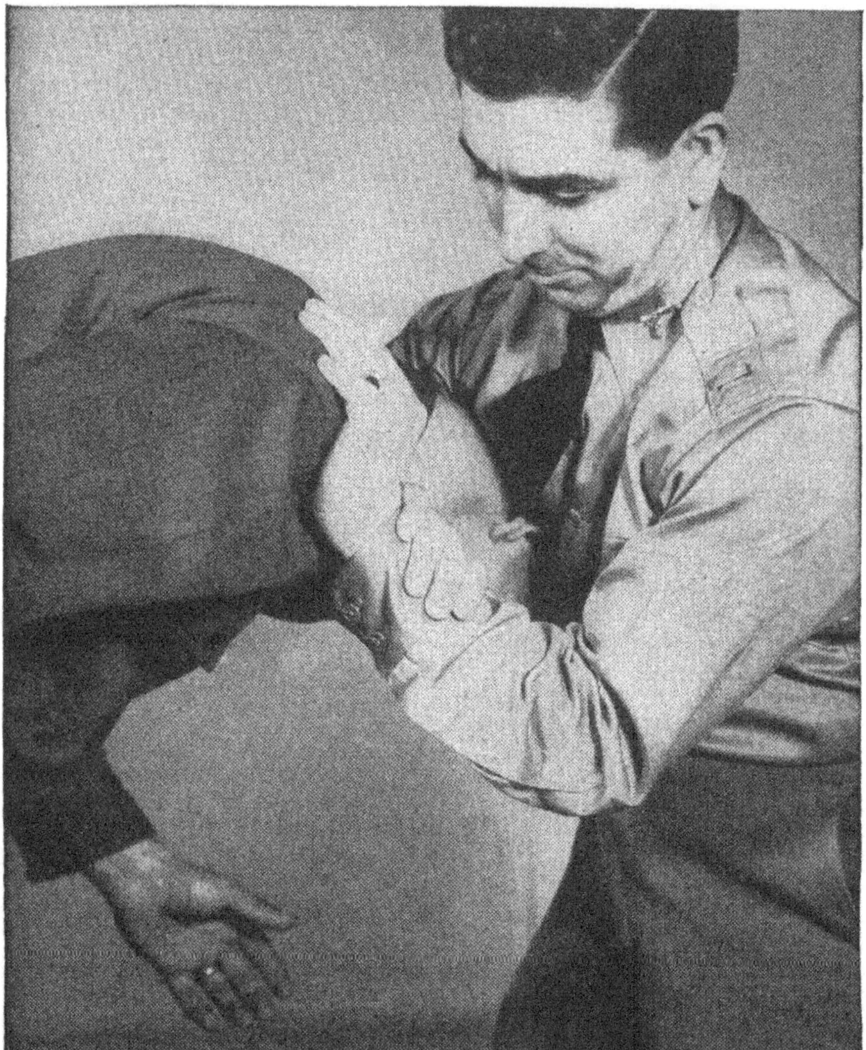

FIGURE 103 ②.

left hand is placed on the back of his head while your right forearm is crossed from the right under his neck (fig. 104 ①). It is further brought to the left and locked on your left forearm (fig. 104 ②). From this position, pressure is brought to bear by pressing with the hand on the back of the head and leaning forward. A quicker and more potent choke from the same position is illustrated in figure 104 ③. Instead of

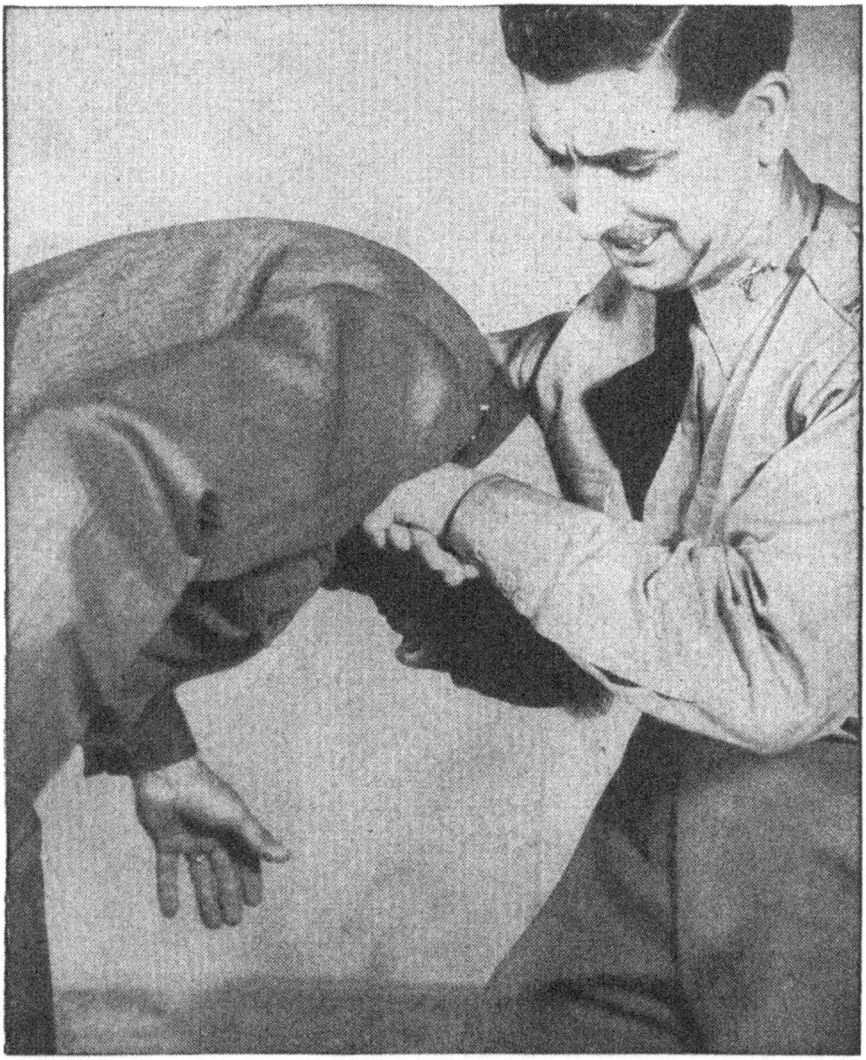

FIGURE 103 ③.

locking the left hand on the back of your victim's head, it grasps the little finger edge of your right hand and is drawn quickly toward your chest as your right shoulder pushes against the victim's head.

■ 109. SUBCLAVIAN NERVE PINCH.—This is a very painful nerve center and is a very good method to use in causing your victim to release an object or a person. The thumb is dug into a point

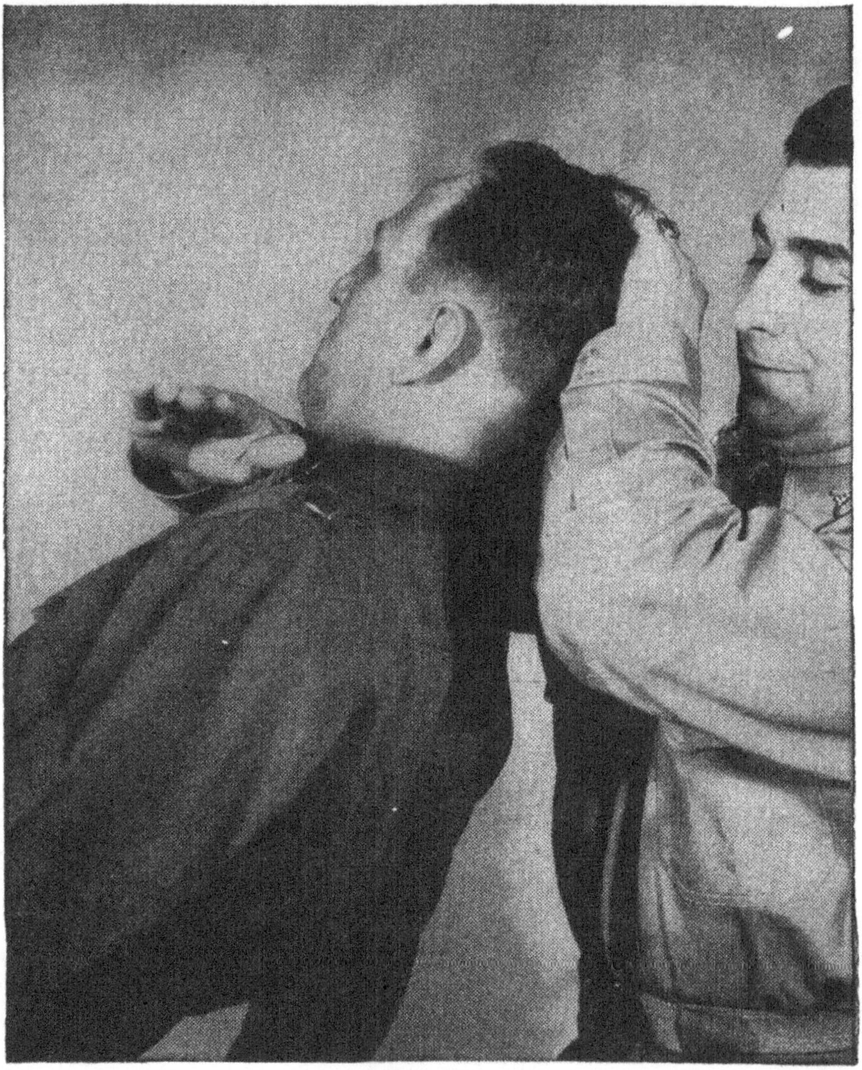

FIGURE 104 ①.

at the base of the neck and just above the clavicle bone (fig. 105).

■ 110. AXILLARY NERVE PINCH.—This nerve pinch has the

same function as the subclavian. It is a very good method of walking an individual inconspicuously out of a room or any place that you will. The thumb or fingers can be brought to

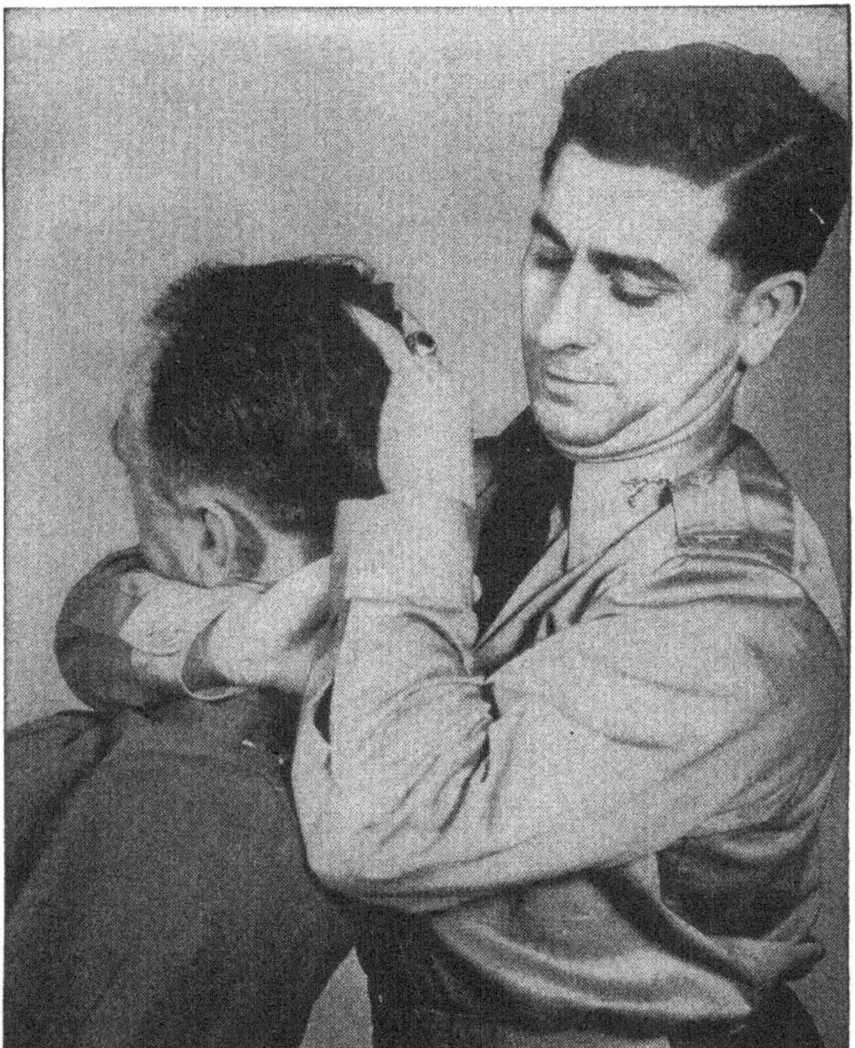

FIGURE 104 ②.

bear on an exposed nerve underneath the armpit and close to the base of the bone in the upper arm (fig. 106).

■ 111. CROSS COLLAR HOLD.—This is a certain method of subduing an individual quickly. To be effective, it must be worked on an individual with an open collar or open lapels on his coat or jacket. The hands are crossed at the wrist and the fingers grasp the collar on the opposite side with

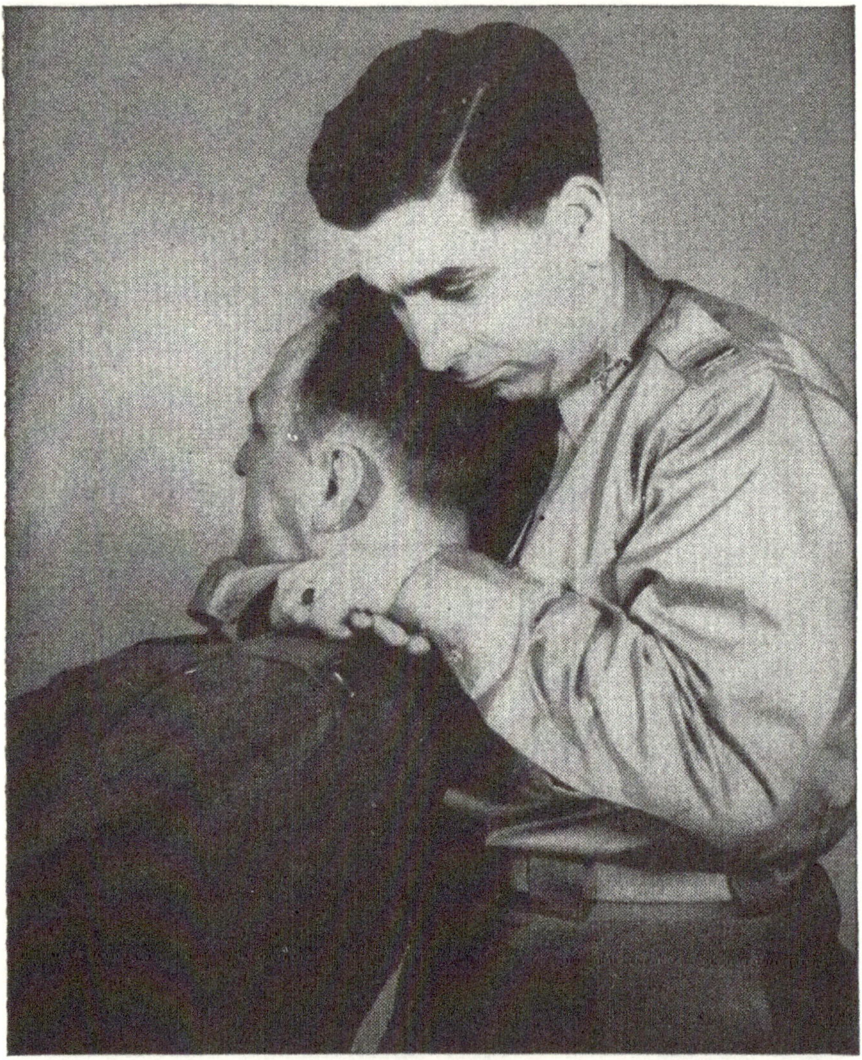

FIGURE 104 ③.

the fingers inside and the thumb on the outside (fig. 107 ①). Grasping strongly with the fingers, the arms are then scissored, causing your opponent to drop to the ground unconscious within a second (fig. 107 ②).

■ 112. **Death Blows.**—The three following blows are to be used only in dire emergency, when it is absolutely essential that your attacker is put out of action permanently:

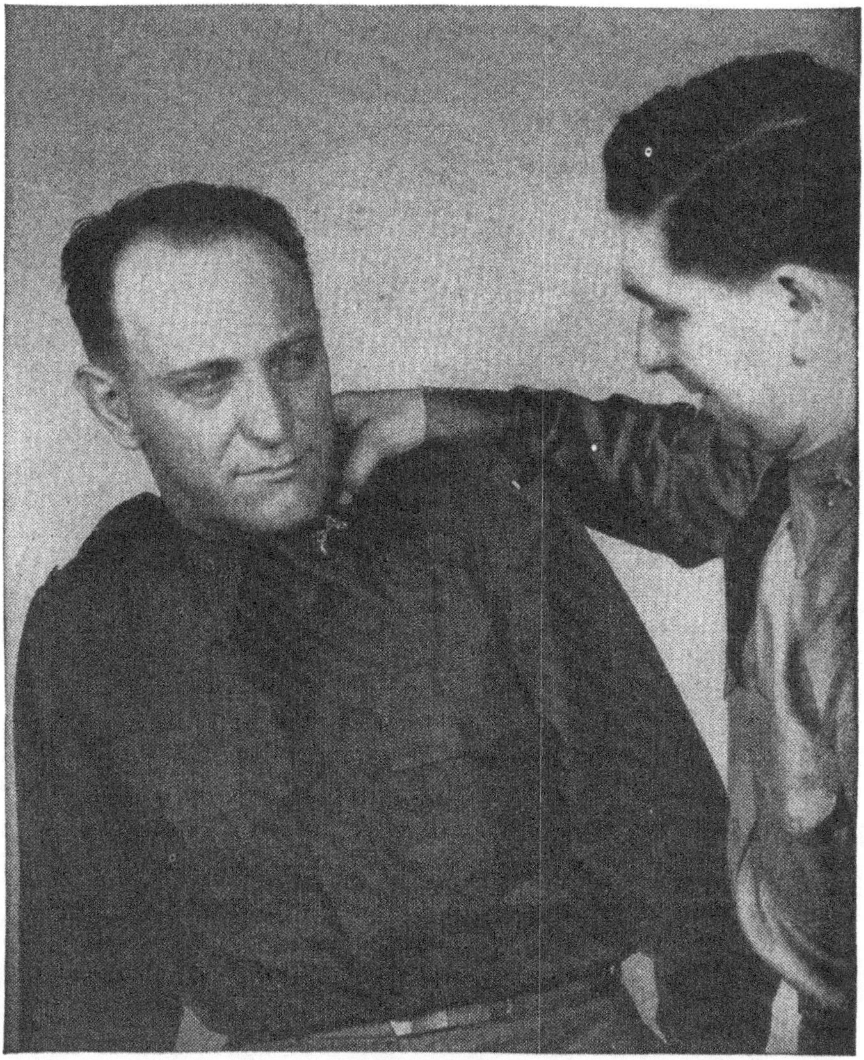

Figure 105.

a. Eye stroke.—A blow struck with the little finger edge of the hand in the manner illustrated in figure 108 will break a paper-thin bone, driving a sliver into the brain and causing instant death.

b. Nose stroke.—A blow struck in the manner illustrated in figure 109 with a great deal of force will cause death; a lighter blow, very severe concussion. The blow is struck with the

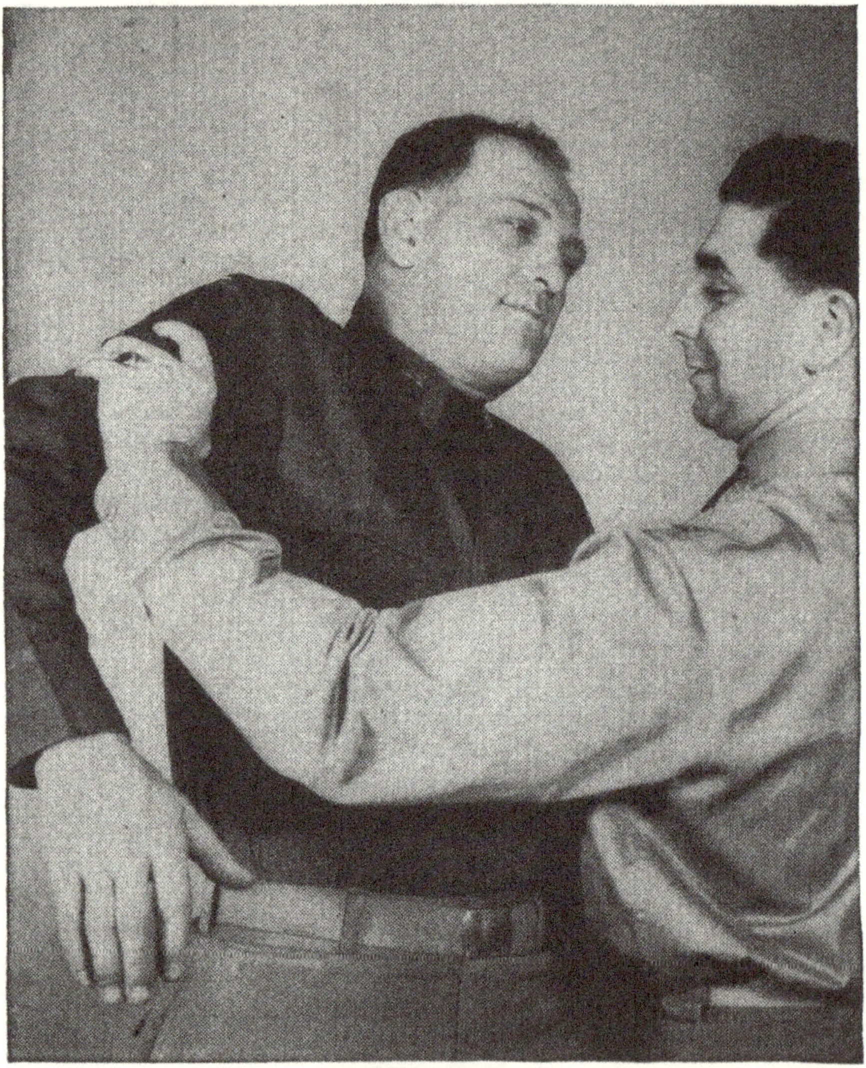

FIGURE 106.

edge of the hand underneath the nose and diagonally toward the ears.

c. Celiac plexus stroke.—The blow illustrated in figure 110, struck just below the base of the breast bone with the fingers held as closely together as possible and with the palm uppermost, will burst the aorta, causing death due to internal hemorrhage.

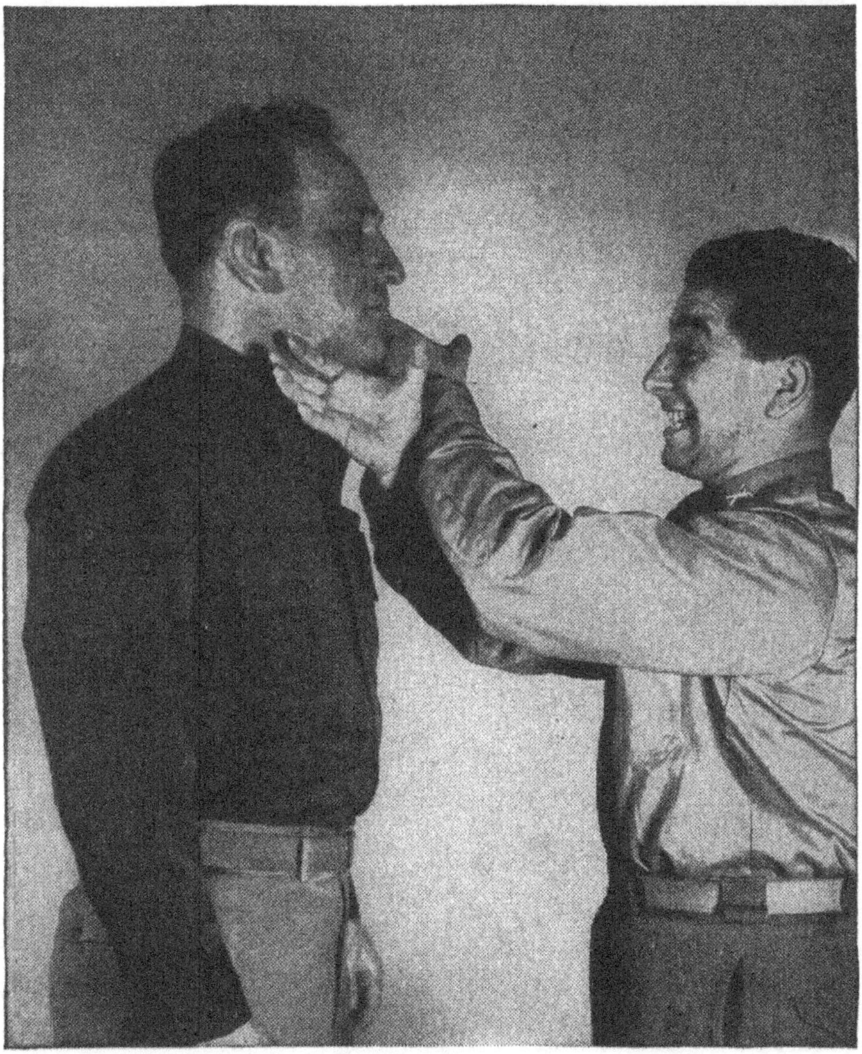

FIGURE 107 ①.

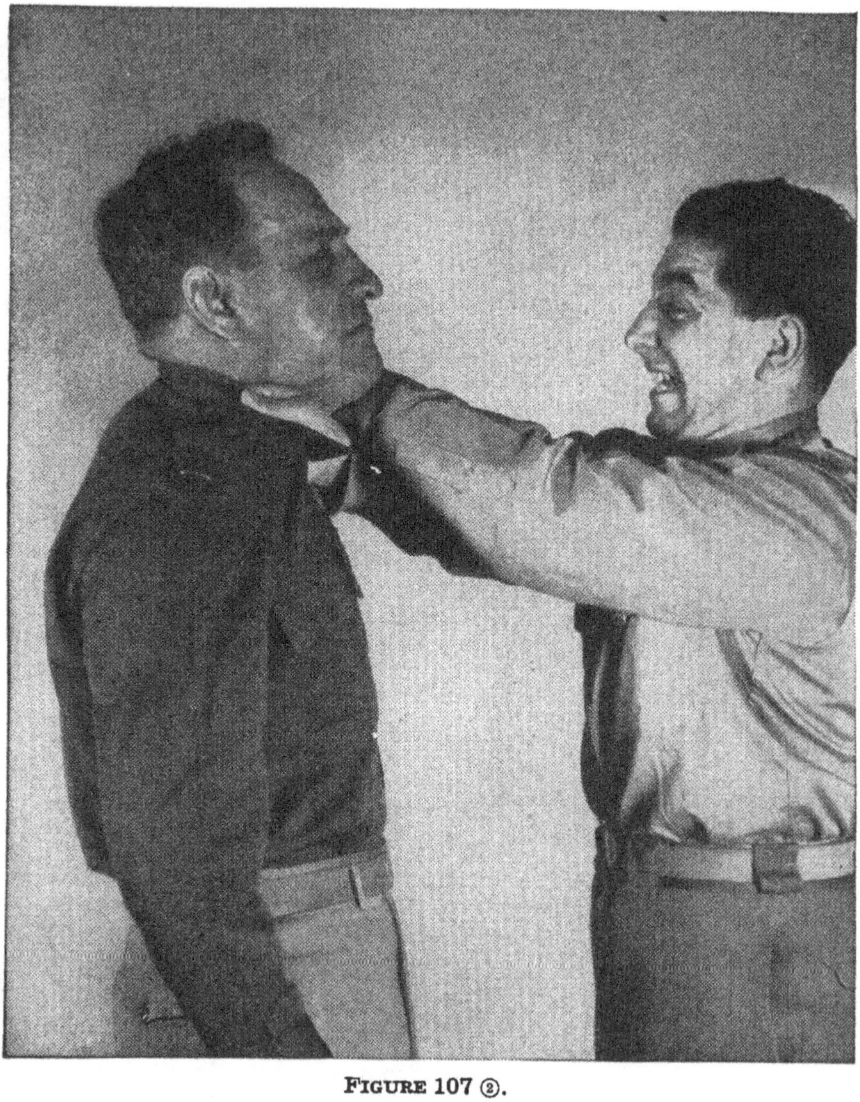

FIGURE 107 ②.

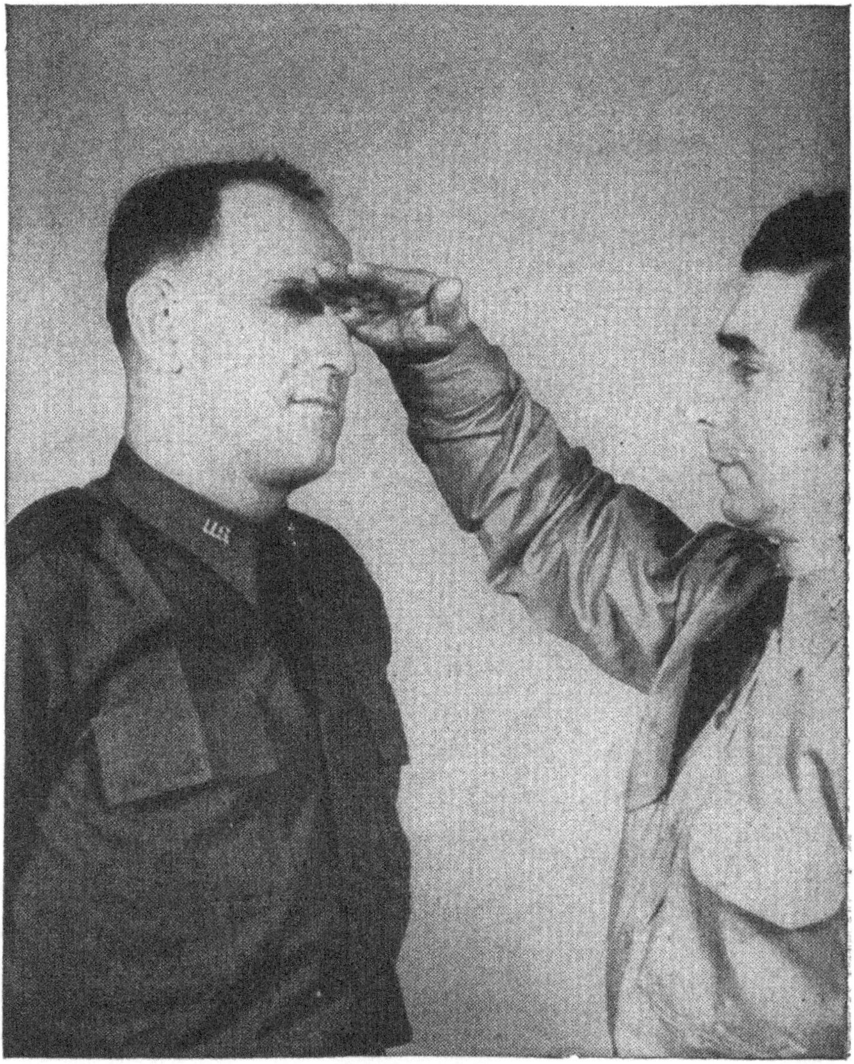

FIGURE 108.

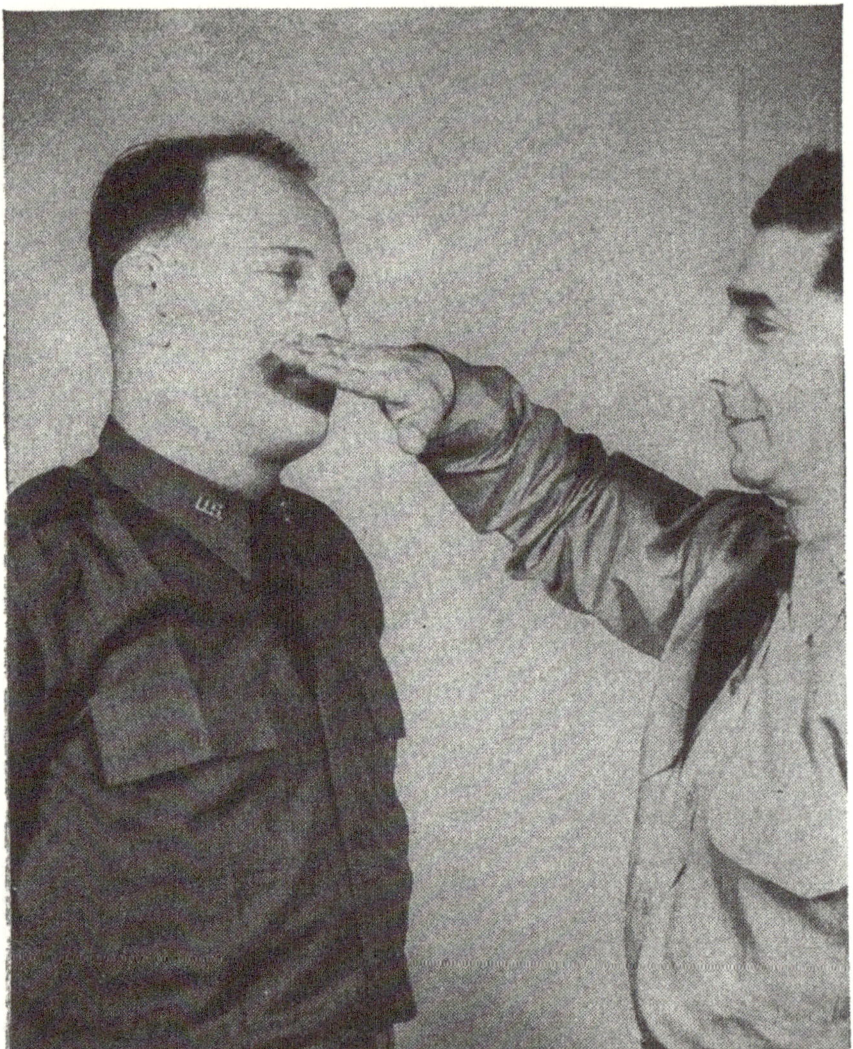

FIGURE 109.

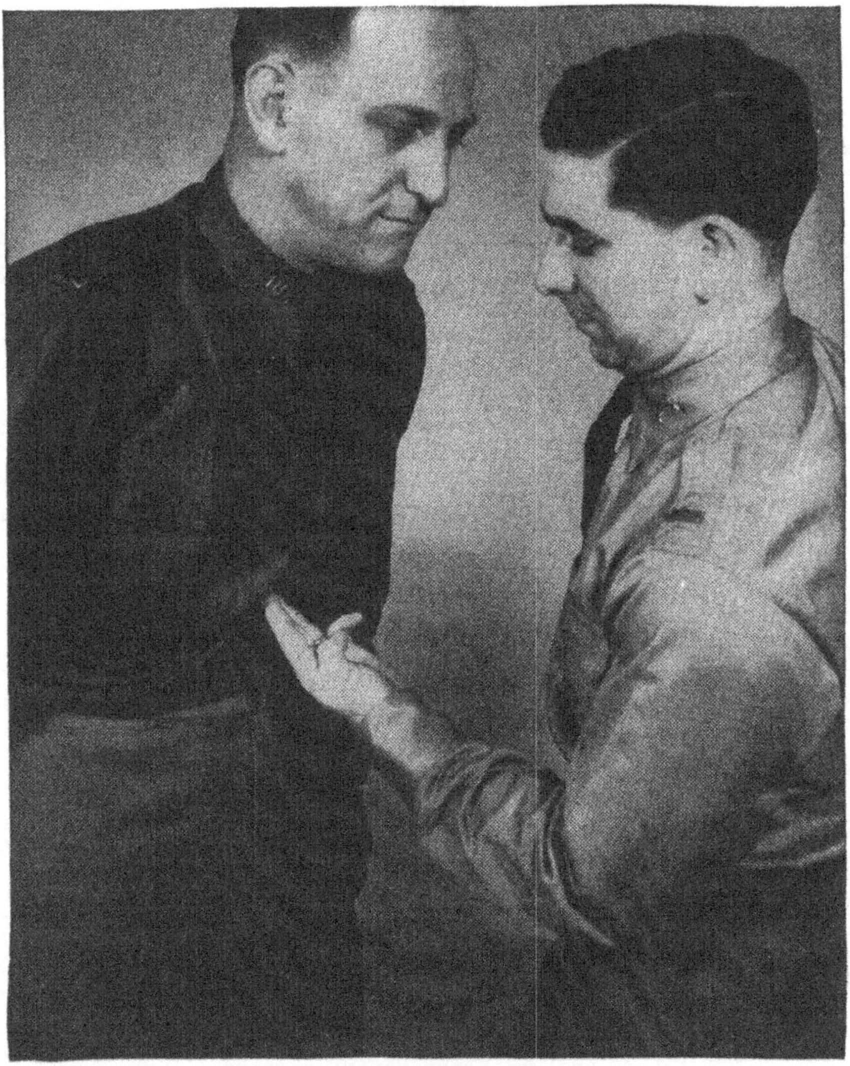

FIGURE 110.

INDEX

Made in the USA
Coppell, TX
03 September 2026

88978436R00187